Berries

Berries

WRITTEN AND ILLUSTRATED BY

Roger Yepsen

W.W. NORTON & COMPANY
NEW YORK LONDON

This book is for my mother
and the blueberries on Lille Ø

Printed in Italy
First Edition

Library of Congress Cataloging-in-Publication Data

Yepsen, Roger B.
Berries / written and illustrated by Roger Yepsen.— 1st ed.
p. cm.
ISBN-13: 978-0-393-06031-7 (hardcover)
ISBN-10: 0-393-06031-4 (hardcover)
1. Berries. 2. Cookery (Berries) I. Title.
TX813.B4Y47 2006
641.6'47—dc22
2006000707

W.W. Norton & Company, 500 Fifth Avenue, New York, NY 10110
www.wwnorton.com
W.W. Norton & Company Ltd., Castle House, 75/76 Wells Street, London, WIT 3QT

1 2 3 4 5 6 7 8 9 0

Contents

BLACK RASPBERRY

CHAPTER 1

Perishable and Precious

"Berry-pickers, like poets, are born, not made." So wrote Martha Bockee Flint in *A Garden of Simples,* published in 1900. Maybe so. Some people will take a walk in the countryside and be oblivious to the berries, while others stand rooted by each shrub, purple-handed, until it has been completely stripped of fruit.

And yet a summer of berry picking can change your sensibilities, because you have to be keenly alive to your surroundings in order to find the berries (and avoid their thorns). It would be interesting to query a sample of environmentalists and naturalists and ask if their choice of profession was influenced by picking berries as children. America's first environmental radical, Henry David Thoreau, noted that wild blueberries were valuable not only as a food but also as an introduction to the joys of the natural world for the New England children who picked them as a family chore.

Berries have this potential power because they are edible jewels, compounded of sunlight and soil and floral perfume. Their colors are so pure and saturated that we call on berry names when defining a range of hues from warm pinks to chilly purples. Berry flavors and aromas are concentrated, too. While some offer an ambrosial sweetness, others are as assertive and individualistic as spices. Strawberries come across candylike and child-pleasing; at the darker end of the berry spectrum, black currants put off a murky reek that bewitches some people and repels others. Because berries have such distinctive flavors and scents, *grape* wines often are described in berry terms. Read the back label on a big red vintage and you may find a highly imaginative blurb claiming the wine is redolent of black currant jam or ripe elderberries—a curious convention, in that relatively few Americans are familiar with either.

This book is a guide to getting more involved with berries, including neglected ones that have all but disappeared from the American diet and garden. Berries can be used any time of day, in any meal or type of dish. It may be impossible to improve on the taste of fresh-picked berries, but you'll find recipes that suggest ways of taking advantage of their color, flavor, and recently confirmed nutritional benefits. These ideas range from cobblers to cocktails, from Black Currant Corn Pancakes in the morning (see page 56) to a glass of homemade Tutti Frutti Liqueur before retiring at night (page 38). Along with familiar tarts and pies, the recipes include such quaint treats such as a buckle, a grunt, a slump, and a fool.

One reason we aren't on more intimate terms with berries is that they can be perversely fragile. Most don't take well to large-scale agriculture and long-haul shipping. Some, such as the yellow raspberry and salmonberry, are almost impossible to transport or store. More than 40 percent of certain fresh berries are lost between harvesting and reaching the consumer's table, which helps to explain their high cost.

BERRIES AS A STAPLE

Across much of the cooler reaches of the Northern Hemisphere, berries once were a basic part of the diet, whether eaten fresh or processed into jams, juices, and alcoholic beverages. Back before the ready availability of citrus fruits, berries provided a crucial source of vitamin C. Native Americans may have liked the taste of strawberries baked in their cornbread (see Pan-Fried Strawberry Polenta, page 139), but the berries were there out of nutritional necessity, not simply as a novel twist on a timeworn recipe. The Senecas developed a combination berry picker's basket and evaporating tray, which held the just-picked berries and also provided a latticework on which the fruits were preserved by drying. Although berries were an important part of the Native American diet, the women and children who did the picking didn't seem to treat the job as onerous. According to period accounts, this was a social event, with much singing as they worked.

As the Europeans settled throughout berry country, children continued to be responsible for much of the harvest. Somewhat uncharacteristically for a man who thrived on solitude, Thoreau mentioned feeling lonely on his daily hikes after the blueberry season had ended and the young people with their buckets were gone from the countryside.

Over time, the berry pickers were more likely to be out in the fields to earn money than to supply their family's needs. Thoreau bemoaned the change, calling it a "desecration" to convert wild blueberries to cash; the tiny fruits had become "enslaved" by capitalism. With this shift, landowners began nailing up no-trespassing signs around their best berry fields, something new in the American landscape. "Nature is under a veil there," Thoreau wrote of the posted acreage. Meanwhile, farming practices were depleting the soil. No longer were entire hills tinted red by wild berries, as the first European settlers had noted. By the end of the century, observers of the landscape were grumbling nostalgically about the berry harvests of the past.

Still, gathering berries has remained a way of life across much of rural North America. People continue to anticipate each flush of berries, from the first strawberries of early summer to the last diehard raspberries of fall. Other activities are postponed as these berry lovers scout the best patches while the fruit still is unripe, then head out with their favorite collecting containers for long, patient hours of foraging. By doing so, they are taking an intimate part in a traditional way of gathering food that goes back thousands of years; there is no better way to step into this continent's pre-Columbian history than to spend an hour picking berries. In a pinch, knowing where to find the berries can save a life. A 1934 newspaper account told how a Nova Scotia woman, lost in the wilds for nine days, was able to survive on the blueberries and raspberries she found. When rescued, she was six pounds lighter but in good health.

Berry lovers are protective of their prized patches. My parents discovered an extensive blueberry thicket along an old logging road within a two-mile walk of their Adirondack summer home. The berries went into pies that they shared with neighbors. And the neighbors, reasonably enough, began asking where all the berries came from. "We could tell you, but then we'd have to kill you," was my parents' standard line. To escape notice, they took to wearing binoculars as a subterfuge while picking. Whenever a car or truck approached, they hid their cups of berries and pretended to be eying a warbler.

In Montana, huckleberries enjoy the same devotion as blueberries do in the Northeast. Ellen Bryson can locate patches of the ripe berries just by catching their scent while driving through the countryside with her husband at the wheel. When she catches a whiff, she yells to her husband to stop. If the people in a following vehicle wonder what's going on, Ellen's license plate gives a clue: it reads HKLBRY.

GONE PICKING

Some of the best harvests are to be had on public land. In the Gifford Pinchot National Forest in Washington State, for example, visitors flock to pick strawberries, salmonberries, thimbleberries, blackberries, elderberries, and, most popular of all, huckleberries. The demand is such that some state and national parks impose regulations on when and where berries may be picked. You may need to get a permit in order to pick what are considered to be commercial quantities of berries. Some areas have been set aside for Native Americans to harvest, allowing them to continue a life-sustaining activity that has gone on for thousands of years. While checking on a park's regulations, ask for tips on where to head for the best picking. In Wisconsin, the Nededah National Wildlife Refuge recognizes the "long-standing historic use" of the park for berry picking, and rangers help to guide visitors to recently burned-over patches of land where berry plants are apt to flourish.

Devoted berry pickers are particular about their containers. In upstate New York, just west of the Adirondack Mountains, Pooniel Bumstead has nearly worn out her traditional Native American berry picking basket. It has a small opening at the top and bellies out below. Ellen Bryson favors a plastic bucket with a lid in which she has cut a flap. If she falls (and she routinely takes a tumble in the underbrush), she might get a scratch but the berries are secure. Both women hang the containers around their necks with a cord to free their hands.

Foraging for berries isn't without its potential dangers because you might run into a bear doing the same. One precaution is to carry a tin can with pebbles in it; at the sound of something large in the bushes, you give a good, noisy shake (if you aren't already shaking). My mother never goes berrying without a large brass whistle around her neck.

Another potential hazard is picking poisonous berries. Be confident in your identification of the wild berries you are after, consulting either a field guide or an experienced berry picker if necessary. If you have young children, be aware that they are apt to follow your example and go picking on their own. To them, the bright berries of deadly

WHIPPED BERRY CREAMS

nightshade are apt to look attractive—and edible. So it was with our son Rhodes, who came into the house one summer day with berries smeared around this mouth and on his hands. When we asked where he had been, he led us to a hedgerow where both nightshade and red raspberries grew. We called the local poison control center and followed their advice—administer an emetic to force the poor kid to throw up. Instruct young children to pick berries only when you are with them, if at all. Scout your property for plants bearing poisonous berries, and either fence them off or prune the berries and dispose of them. These include red and white baneberry, burning bush, chokecherry, lily of the valley, mayapple, oleander, and pokeberry. Keep an emetic (such as syrup of ipecac) on hand, along with the number of your local poison control center.

Easier than foraging is to pay to pick your own on a farm that sets out bushes for harvesting by the public. While cultivated berries may not necessarily have the full flavor of wild ones, the picking goes a lot faster and you are apt to find a range of varieties of a particular berry, each with its own qualities and ripening season.

THEY'RE GOOD FOR YOU, TOO

Berries have been getting a lot of attention from nutritionists and medical researchers. They may contain impressive levels of vitamin C, potassium, and antioxidants credited with benefits ranging from reducing cancer risk to resisting the effects of aging. According to an Agricultural Research Service study of fruits and vegetables, blueberries scored highest in antioxidants. (Oxidative processes in the body have been associated with causing both certain cancers and heart disease.) There even is evidence that blueberries contain nutrients which may improve mental functioning, such as the powers of concentration and memory. A researcher at the USDA Human Nutrition Research Center on Aging at Tufts University suggests that a diet consisting of 1 or 2 percent blueberries "may reverse short-term memory loss and improve motor skills."

Some of the responsible nutrients are right before your eyes—anthocyanins are antioxidants responsible for creating the inky colors of darker berries (as well as the "blood" in blood oranges). Berries also deliver pectin, which is thought to help lower levels of the type of cholesterol linked with cardiovascular disease.

The role of fruits in the diet has taken on new prominence, now that federal guidelines recommend eating *nine* servings of fruit and vegetables each day, up from five—at a time when most Americans are getting only three. For an idea of what this requirement looks like on a plate, that's about four and a half cups of these foods per day.

So, berries aren't just for desserts and occasional picking while on a walk. They can figure as a key part of your diet, fresh or cooked or processed.

GROWING THEM

The most dependable source of berries, and the freshest, is the home landscape. Unlike many foods that could be called delicacies, berries grow with a minimum of fuss. If you've succeeded with zucchini, you'll probably manage with berries.

Most berries are relatively self-sufficient. The plants establish themselves quickly, require a modest amount of annual maintenance, and can be expected to yield dependably for years. Try a couple of easygoing blueberry bushes, with their compact form and fall color, and discover why berry plants once were fixtures in North American backyards. Currants, gooseberries, and elderberries also are mild-mannered and attractive. If you are in need of a ground cover, look into lowbush blueberries, lingonberries, and tiny checkerberries with their wintergreen-flavored leaves.

Strawberries are a delectable exception. Although wild strawberries seem to flourish just about anywhere without anyone's prompting or primping, a bed of a plump domesticated variety will need attention if it is to produce well. Nor does a mat of strawberries add much to the landscape. Raspberries and blackberries offer their own challenges, rapidly growing into briar patches if unchecked. A sprawling, briared bed of berries isn't ideal for a cramped yard where young children play.

PLANNING THE BERRY PATCH

Before buying berry bushes, make sure that you are clear on what these plants will demand of you in terms of pruning, staking, and trellising. If you are looking at the offerings of a mail-order nursery rather than buying locally, pay particular attention to the numbered hardiness zones stated for each variety; if you are in zone 5, as indicated by the maps printed in catalogs and gardening books, a berry plant hardy to zone 6 will be at risk each winter.

Blueberries, currants, gooseberries, and elderberries are among those berry plants that can stand alone without being propped up. They are good choices if you don't care to get involved with rigging up a support system. Raspberries, blackberries, and their relatives grow best if their wandering, drooping canes are given some rigor and order. A simple alternative is to drive a wooden or metal stake in the ground by each plant and tie the canes to it. Another approach is to buy or make the sort of wire cages used to help tomato plants reach for the sun. If you have several plants that need support, it may be more efficient to handle a row of them with a trellis, which looks like a stunted clothesline. Two or more wires are stretched taut between end posts. That sounds simple enough, but you'll have to dig holes for the posts.

HANDLE WITH CARE

Fresh berries generally are evanescent things, sensitive to bruising and mildew. Promptly discard any berries that appear damaged, overripe, or moldy—one bad berry can spoil the whole basket. If berries are wet, allow them to dry on paper toweling. Place them in a container lined with more paper towels and keep it in coldest part of the refrigerator.

BERRY LIFE SPANS

Here are estimates of how long various berries will remain in good shape after picking if refrigerated. Note that store-bought berries will have a shorter shelf life, depending on how they have been handled.

Raspberries	Up to 5 days
Strawberries	Up to 1 week
Gooseberries	Up to 1 week
Blueberries	Up to 2 weeks
Cranberries	1 month

The ideal temperature for most berries is just above freezing. If you have a lot of berries on hand, you may want to adjust your refrigerator's thermostat so that the coldest part of it hovers around 33 or 34 degrees Fahrenheit. As a general rule of thumb, each hour that picked berries spend at room temperature means one less day of shelf life. Delay rinsing berries until just before you use them.

For storage over a longer time, spread a layer of berries over a baking sheet and freeze them in batches, putting each batch in a freezer bag or other container as you go. Keep them in the freezer for up to six to twelve months.

Freezing changes the structure of berries to varying degrees. Strawberries turn pulpy because of their high water content; blueberries and cranberries, on the other hand, retain their shape. As the water within the berry's cells turns to ice, the cell walls are broken, releasing juices and softening the texture—an advantage if you will be converting the berries to juice.

BERRIES IN THE KITCHEN

While it may be true that no one has ever come up with a recipe that improves upon a fresh strawberry or raspberry, cooking with berries is fun—and healthful, because you have more ways of incorporating them in your diet. A good recipe also changes the flavor and texture of berries in interesting ways. With heat, acids are toned down and the berries may become thick and jamlike. Then, too, certain ingredients seem to bring out the best in a berry. Nutmeg is a natural partner for the herbal tang of blueberries, rhubarb fills in any culinary shortcomings a strawberry might have, and mulberries can use the help of lemon and other fruits.

For some of us, berry recipes are part of a long family tradition. On one side of my family, a great-grandmother learned to make berry desserts as a girl on Denmark's Jutland peninsula. On the other side, my grandfather grew a pick-your-own strawberry field and a great-grandmother made pies from the elderberry bushes in the backyard of her Indiana home. Ask around in your own family, and you may find an intriguing berry recipe that has been handed down over the generations.

Old recipes are interesting for what they suggest about how people's lives have

BERRYMETRICS

Berries vary a lot in how much a given volume measure will weigh. A cup of little wild strawberries will weigh more than the three or four mammoth strawberries a cup can hold. The yield of juice is variable, too. If you filter out all of the pulp to make a crystal-clear juice, you'll end up with a reduced amount.

1 pound berries equals about 2½ to 3 cups
1 quart berries yields about 1 pint strained juice

changed over the years. For example, early cookbooks assumed that readers would have ready access to various fresh berries—and the discretionary time needed to pick them. In an edition of *The Joy of Cooking* from the 1930s, Irma S. Rombauer's recipe for Gooseberry Conserve asks you to get your hands on four quarts of a berry seldom seen in supermarkets or roadside stands. Since World War II, berry plantings are no longer a standard part of American yards, having gone the way of hand pumps and chicken coops. The change was reflected in Beatrice Trum Hunter's pioneering *The Natural Foods Cookbook,* published in 1961: the index lists far more recipes under *Brains* than under *Berries.*

Today, we are enjoying a berry revival, for a couple of reasons. Some credit goes to the widely publicized reports of the healthful compounds found in berries. You can expect this news to be reflected in a greater selection of berries and berry products. Mail-order nurseries have been quick to point out berry varieties that are especially rich in antioxidants, vitamins, and minerals. A second factor is that supermarkets and even mom-and-pop stores now offer fresh berries flown in from wherever on the globe summer happens to be taking place. Frozen berries are readily available, too, and at a more reasonable price.

JUICE, JAMS, AND JELLIES

Most berries have a short shelf life. You can do as people have done for centuries and capture the berries' summer essence by putting up clear juice and jellies, or rich preserves, butters, and even berry leather. These processes discourage spoilage by microorganisms by concentrating the sugar content, lowering the moisture level, or both.

Preparing those colorful jars is agreeable work and completely safe if you are mindful of the directions. You can have the fun of coming up with your own blends: raspberry and red currant often are paired in northern European recipes; and strawberry works well with habanero peppers. Expect your homemade jellies and jams to taste better than supermarket fare. Commercial red currant jelly, to take an example, may be so bland that you can identify it only by looking at the label. Even the folksy-looking jams and jellies sold at roadside stands and general stores may be private labelings of mass-produced goods shipped in from a distant factory.

From one berry to the next, the directions for processing are pretty much the same. *Jellies* are made from pure juice. It's up to you whether or not to include the pulp that constitutes a *jam*. *Preserves* are chunkier still, containing identifiable berries or pieces of berry. Berry *leathers* traditionally were made by drying purees in the sun, and an oven will do the same. Add sugar or another sweetener to berry juice and you have *berry syrup*.

Berries reach their richest flavor and best color when fully ripe, but some home cooks prefer to preserve slightly less-than-ripe fruit because its higher level of pectin can result in a firmer set; you might experiment with a blend of ripe and underripe berries. This mix also can add piquancy to a pie. Wild berries take longer to find and pick than cultivated berries, but they may be more flavorful. Don't pick berries from along roadsides. The fruit may be contaminated with the lead that is the legacy of gasoline additives, or rendered unfit by the spraying of herbicides. In any event, you can count on roadside berries to be grimy, and berries don't hold up to a good scrubbing.

BERRY JUICE

Berry juices are enjoyable in their own right, and they are the first step in producing jewel-clear jellies, wine, and many recipes.

Begin by pureeing the berries, with a potato masher or a food processor. A brilliantly clear juice can be had by using either a jelly bag or cheesecloth draped inside a colander to strain this puree. For maximum clarity you may choose to allow the juice to drip on its own rather than prodding, poking, or squeezing the cloth. Small amounts of juice can be poured though a coffee filter. You also can use a strainer, so that some pulp passes through for a richer but cloudier juice. The yield will be greater if you apply pressure with the back of a large spoon.

To coax more juice from berries, you can use either heat or cold. Simmering the berries in a bit of water will break them down and make them juicier, as will freezing them and allowing them to thaw. Another heat-based method is to use an electric or stove-top steam juicer. The juice is collected in a container within the juicer. High-quality steamers of stainless steel can be expensive, but they spare you the mess of using cheesecloth and their convenience is a big asset if you process a lot of berries.

You can *buy* the berry juice, of course, from a supermarket or natural foods store. It may be more economical to buy gallon cans of berries with juice from mail-order companies that sell principally to winemakers. A search of the Internet will turn up a good variety, including loganberry, boysenberry, marionberry, elderberry, gooseberry, and black currant. This source gives you the chance to explore the flavor of berries from outside of your region, not only in making juice, jams, and jellies, but in any recipe that doesn't require fresh fruit.

Refrigerate fresh juice for up to a week. To store juice for longer periods, either preserve it in a bath of boiling water as described in the following recipe for jelly, or freeze it. Another traditional means of preserving juice is to add sugar and let yeast convert the sweet mixture to wine.

Gooseberries and currants contain enough pectin to make a reasonably firm jelly. For other berries, recipes call for either adding pectin or cooking until the mixture reaches a temperature that will cause the sugary liquid to turn firm when cooled.

One way to add pectin to a berry jelly is to include pectin-rich apples in the recipe. But for a full, rich berry flavor, you can make an all-berry jam by adding commercial pectin that has been concentrated from other fruits. There are two sorts of pectin. The kind most widely available at supermarkets requires the addition of a substantial amount of sugar in order to fully activate the gelling process. For example, you might be directed to use 7 cups of sugar with 2 quarts of strawberries. The sweetener isn't there because ripe strawberries need the help, but because that proportion of sugar will result in a firm set. Use less sugar, and you run the risk of producing a runny jelly.

An alternative is to use a *low-methoxl* pectin that gels in the presence of calcium, not sugar. The calcium is supplied in the form of an additive, such as monocalcium phosphate, which comes packaged with the pectin. Using low-methoxl pectin, you need just ¾ cup to 2 cups of sugar for 2 quarts of strawberries. Honey can be substituted for some of the sugar if you are using this type of pectin. If you can't find it at the supermarket, try a natural foods store or a mail-order source.

Gather canning jars or reused glass containers and check for cracks and chips as you wash them. Sterilize the jars in a canner or a large pot of boiling water for 10 minutes. (At higher altitudes, water boils at lower temperatures so that longer processing times are needed; add at least 1 minute for each 1,000 feet you are above sea level.) Keep the jars in the hot water until you are ready to fill them.

Place the juice in a pot. Measure the pectin as directed by the instructions that come with the product and stir it into the juice. Bring the juice to a rolling boil (the point at which the boiling cannot be quieted with a stir). Add the sugar. If the juice seems insipid, add lemon juice or red currant juice; berry recipes may include tart ingredients—juice, or an acid—to help the mixture to set. Stirring constantly, return the juice to a rolling boil for 1 minute. Skim any foam that has formed. Remove from the heat.

Take out each jar in turn from the hot water and place it on a pie rack or dry dish towel to prevent it from cracking. Ladle the juice into the hot jar to within ¼ to ½ inch of the top. Put the lid tightly in place.

Boil the sealed jars in the canner or pot for at least 10 minutes, again adding 1 minute for each 1,000 feet of altitude. Remove the jars from the water bath. If all goes well, gelling will take place as the jars cool (and may continue for a week or more). Check to make sure that there is a good vacuum seal on all of the jars. On some metal lids, you can perform a simple test by pressing down on the center; it will click downward and stay there if a vacuum is present. A glass lid with a rubber ring should resist coming off with a gentle pull. If either test fails, refrigerate the jelly and use it soon.

To make jam, you can use either straight pureed berries or a mixture of puree and juice. For preserves, reserve some whole berries (or slices of large strawberries) and stir them into the puree. Proceed as described for making jelly.

NO-PECTIN JAM

You can make jam or jelly without pectin if you add sugar and cook the sweetened juice to a temperature that signifies the mixture has reached the necessary concentration of sugar. This is a little tricky and it can be messy as well. If you err on the side of too cool, the jelly or jam will be runny. If you allow the temperature to climb too high (and it can do so quickly), you're apt to end up with a pan lined with a treacly, hard sheet. The surest way to determine that you've hit the mark is to use a candy thermometer so that you can take the jelly or jam off the stove precisely when you've arrived at the temperature indicated for your altitude. At sea level, you stop boiling when the thermometer reads between 217 and 221 degrees F; lower this range 2 degrees for every 500 feet of altitude.

If you don't have a candy thermometer, the old-time standby is to perform the plate test. Chill a plate in the refrigerator. Drop a scant teaspoon of the cooking mixture on the plate. If you see a ring of moisture around the blob, and the blob runs when you tilt the plate, the jelly needs more time. As soon as the sample cools to the consistency you're after, take the pot from the stove.

FREEZER JELLY AND JAM

You can use freezing temperatures instead of boiling to inactivate food-spoiling organisms. Prepare the pectin by stirring it in boiling water and then pouring into either berry juice or puree. Stir the mixture, allow it to cool, and pour it into freezer containers. Leave the jelly or jam at room temperature overnight in order to set, then place it in the freezer. After being thawed for use, the product should be good for at least two weeks if kept refrigerated.

BERRY SYRUP

With its added sweetener and thicker consistency, berry syrup can be used as a topping for pancakes, oatmeal, or ice cream; as a spritzer (see the following recipe) or a cocktail ingredient; and in recipes when fresh or frozen berries aren't available.

Makes about ½ pint.

2 pints fresh or frozen berries
1 cup water
2 tablespoons lemon juice (optional)
1 cup sugar

Prepare berry juice as described previously, retaining some of the pulp if you don't care to have an absolutely transparent syrup. Stir in the sugar over low heat until dissolved. For a concentrated syrup, simmer with the lid off. Add the lemon juice if you think the syrup would benefit from it. Freeze or refrigerate the syrup, or use the water-bath method for longer storage.

BERRY LEATHER

For centuries, fruit has been preserved by dehydration. Native Americans made a berry leather of strawberries and raspberries by placing the mashed fruit on stone slabs. Properly made berry leather has too little moisture for mold organisms to cause spoilage, while remaining pliable—and fun to eat. You can make leather from a single type of berry or combine a few berries and other fruits.

Line one or more cookie sheets with plastic wrap. To keep the plastic in place, tape it around the edges. Puree the fruit in a blender. For a sweeter product, add 1 or more tablespoons of sugar or honey per cup of fruit. Strain the puree if it is very seedy. Pour the pureed fruit onto the cookie sheet and spread it evenly with a spatula, leaving some plastic showing around the perimeter so that you'll be able to peel off the finished leather. Use additional cookie sheets as necessary to produce a relatively even layer of no more than ½ inch.

The old-time, zero-kilowatt method for dehydrating fruit is to set the trays out in the sun with some sort of permeable covering to keep out the bugs, such as window screens or a tent of cheesecloth. Not surprisingly, this works best in a sunny, dry climate. Lacking that, you can use the oven. Set the temperature at 145 degrees F. or the "warm" setting, and keep the door open a few inches to vent the moisture. Drying times run to five hours or more, depending on your oven and the moisture level of the fruit. If the leather feels mushy, there may be enough moisture to promote spoilage, so allow more time for drying. Peel the leather from the plastic wrap, then place in on a clean piece of wrap and roll it up for storage.

BERRY WINE

Left on its own, berry juice has an inclination to turn into a mildly alcoholic wine. Wild yeasts consume sugars in the juice, producing a fairly low level of alcohol and carbon dioxide bubbles in the process. Yeasts occur naturally on fruit—they are the bloom on the blueberry. To make a wine with a conventional percentage of alcohol, you'll have to add sugar or another sweetener in order to give the yeast more to feed on.

The results of allowing these natural processes to have their way are highly variable, ranging from a splendid wine to something unspeakably vile. In recent times, home winemakers have brought order to their craft by using special strains of yeast under highly hygienic conditions. Consult a book on winemaking or the instructions that come with winemaking supplies to learn the fine points of yeasts and disinfectants.

COOKING WITH BERRIES

Have you noticed that berry recipes tend to lean heavily on sugar, butter, cream, and eggs? Used in quantity, these ingredients act as a safety net that ensures a rich and flavorful result. Feel free to improvise in coming up with lighter versions of recipes that intrigue you. You also can experiment with ingredients that traditionally have been used to complement berries. Keep a supply of oranges and lemons on hand for both their grated zest and juice. Be aware that citrus skins may be treated with preservatives, wax, and colorants; unless you buy organic fruit or pick your own, scrub oranges and lemons with soap and warm water and rinse well before making zest. Once you've deprived an orange or lemon of its protective outer rind, use the fruit promptly or it will turn hard and unappealing.

Vanilla extract figures in countless berry recipes. You might try using vanilla beans instead for a fuller flavor. Or, if you're accustomed to inexpensive supermarket extracts, you may be pleasantly surprised by the seductive fragrance of better brands.

Another way of rounding off the flavor of berries is to add a tablespoon or two of a liqueur, either your own or a commercial bottling. Many recipes call on two French products flavored with bitter orange, Cointreau and Grand Marnier. For a cheaper alternative with about half the alcohol, you can substitute a domestic triple sec; perk it up with orange zest and reduce the recipe's amount of sugar. Other choices are crème de cassis (from black currants), crème de fraise (strawberries), crème de frambois (red raspberries), and Chambord (black raspberries). If you don't have access to these liqueurs or would rather not incorporate alcohol into recipes, you can use commercial or homemade berry syrups instead. When cooking with berries I like to splash a bit of Lillet Blanc, an aperitif that doesn't have the stewy herbal character typical of dry vermouth.

BERRY PANCAKES

Having once flipped thousands of flapjacks as a short-order cook, I've arrived at certain inflexible ideas on how to *not* go about preparing them. First, it's not necessary to lard up the batter with lots of whole eggs and oil and melted butter. Given a good nonstick pan and bit of baking soda, you can skip these traditional ingredients if you choose.

Second, there is the persistent notion that the batter must be lumpy, with dry flour bombs that burst into dust when you take a bite. If the batter is used reasonably soon, don't be afraid to mix those lumps into oblivion. Just don't stir any longer than necessary.

Third, beware of a thick batter. It will only get thicker as moisture is absorbed by the dry ingredients. And a sludgelike batter makes cakes so fat and dry that they lodge in the throat unless lubricated with butter and syrup. (I once visited a flyblown pancake house that offered all the cakes you could eat for an unbelievably low price. The gimmick? The server stopped bringing butter and syrup after the first stack of three.)

As a final sin, several batches of cakes will be piled up like cordwood so that everyone, the cook included, can dine together—on chilled, clammy, leathery discs from which the fragrance has vanished.

This recipe is for single large pancakes, to be served immediately. The cook selflessly turns out one perfect cake for each person before making the last for him- or herself.

As a basic guideline, use 1½ cups of larger berries per cup of flour, or somewhat less for wild blueberries and currants. Blueberries are best dropped onto the cakes as they are ladled onto the pan because their juice can turn the batter a somber shade. Raspberries, on the other hand, will lend a pretty tint and they can be added along with the other ingredients.

Serves four (one large pancake each).

2 cups milk
1 egg
½ teaspoon vanilla extract
zest of 1 orange
1¼ cup unbleached white flour
2 teaspoons baking soda
1 tablespoon sugar
1 teaspoon salt
1 to 1½ cups berries

Whisk together the milk, egg, vanilla extract, and orange zest. In another mixing bowl, throughly mix the flour, baking soda, sugar, and salt. Stir the liquid mixture into the dry mixture until there are no visible flour pockets. Either lightly stir in the berries or sprinkle them on the cakes as soon as they are ladled into the pan.

To make a single large cake, pour a quarter of the batter into a nonstick pan at least 10 inches in diameter, over low to medium heat. Flip the cake when the air bubbles begin to turn firm. Continue baking until the underside is golden brown. Serve immediately with syrup and butter or the toppings of your choice.

VEGAN PANCAKES WITH SMOKY TEMPEH

Light, delicious pancakes can be made without milk or eggs, as is demonstrated each morning at the Black River Café in Oberlin, Ohio. The popular spot is just a short walk from Oberlin College, with its sizable subset of vegans, and the menu offers breakfast favorites without meat or dairy. Owner Joe Waltzer prepares big batches of these vegan pancakes—his recipe calls for 50 pounds of flour—by preparing separate dry and liquid mixes. The two are stirred together as orders come in. Joe explains that this makes the most of the leavening properties of the baking power. As you probably have observed, pancake batter loses its stuff if you allow it to sit around for long. So, if you are making breakfast for a big crew, or if you expect late risers to show up at the table just before noon, consider doing the same.

The Black River Café serves smoke-flavored tempeh as a convincingly bacon-like accompaniment to the cakes. You won't mistake the tempeh strips in this recipe for "the other white meat," but they have a satisfying crunch and excellent flavor.

Serves four.

Dry mix:

1½ cups unbleached flour
2 tablespoons sugar
1½ teaspoons baking soda
1 teaspoon salt

Wet mix:

¾ cup soymilk

½ cup water

1 tablespoon arrowroot powder

2 tablespoons canola oil

Smoky tempeh:

olive oil

16 ounces tempeh

½ teaspoon hickory smoke flavoring

1 tablespoon water

Place the dry ingredients in a bowl and mix well. Place the wet mix ingredients in another bowl and whisk them until frothy. If you will be serving everyone at once, combine dry and wet ingredients, stirring just enough to mix them. Or, if folks are straggling in for breakfast, you might combine just a third of both mixes at a time, for example.

If the tempeh is frozen, either allow it to thaw beforehand or steam it in a covered skillet with a bit of water. Place each tempeh block flat on the countertop and carefully slice horizontally to make two thin blocks. If the blocks are wide, they also can be sliced in half so that they approximate the size of bacon strips. Brown the tempeh on both sides in a skillet with olive oil. As the strips cook, drizzle over them a mixture of the hickory smoke flavoring diluted in 1 tablespoon of water.

OVEN PANCAKE

Here is a way to get around the logistical hassle of frying a mess of pancakes while pouring juice and brewing coffee and so on. This single pancake comes out of the oven hot, puffy, and fragrant.

Serves four.

¾ cup unbleached white flour

¾ cup milk

3 eggs: 1 whole egg and whites of 2 eggs

½ teaspoon vanilla extract

⅛ teaspoon salt
1 tablespoon butter
½ cup berries

Place a cast-iron pan of about 12 inches in diameter in the oven and preheat to 450 degrees F. In a mixing bowl, combine the flour, milk, eggs, vanilla, and salt. Beat the mixture until smooth. Take the hot pan out of the oven and coat the bottom and sides with the butter. Pour in the mixture and spread the berries over the top. Bake for 15 minutes, then reduce the heat to 350 degrees and continue baking for 8 minutes or until light brown. Take the pan from the oven, slice the pancake into four wedges, and serve immediately with syrup or your favorite toppings.

BERRY FRENCH TOAST

I first enjoyed this variation on French toast at the Omega Institute, a new-age center in eastern New York State. As with the previous recipe, this one does splendidly without dairy or eggs. Omega's version uses orange juice; my variation replaces it with berry juice. Either use store-bought juice or make your own.

To prepare 1 cup of juice, place 2 cups of berries in a saucepan containing ½ inch of water. Simmer the berries for 5 minutes, crushing them with a potato masher to help release the juice. Strain through a sieve, pressing with the back of a large spoon to express as much juice and pulp as possible. Stir the sugar into the warm juice. Or, if you are using prepared juice that has been sweetened, you can do without the sugar in the list of ingredients. Because the tahini has a considerable oil content, there is no need to use oil or butter so long as you are frying in a nonstick pan.

Makes six slices.
1 cup berry juice (plus ½ cup berries for garnish, optional)
¼ cup sugar (if juice is unsweetened)
8 tablespoons tahini (sesame seed butter)
1 teaspoon vanilla extract
6 slices of bread

Whisk together the batter ingredients. Dip each slice on both sides, long enough to allow some batter to permeate the bread. Cook over low-to-medium heat in a nonstick pan. Serve with syrup and, if you wish, a dairy-free butter substitute. Garnish with whole berries or sliced strawberries.

BERRY CLAFOUTI

Clafoutis are French desserts that originated in the cherry-growing region of Limousin, and traditionally cherries are the fruit of choice in preparing them. This is a three-layer production, but don't let that put you off: you simply pour a portion of the batter in a buttered baking dish, give this layer a few minutes to set in the oven, then add a layer of berries and top off with the rest of the batter.

Serves six to eight.

1 tablespoon butter
⅔ cup unbleached white flour
½ teaspoon salt
3 tablespoons sugar
4 eggs: 1 whole egg and 3 whites
½ cup milk
2 tablespoon Cointreau or Grand Marnier
½ teaspoon vanilla extract
2 cups berries
¼ cup confectioners' sugar

Preheat the oven to 375 degrees F. Spread the butter around an 8- to 10-inch tart dish or other shallow baking dish. In a mixing bowl, whisk together the flour, salt, and sugar. Whisk in the whole egg, egg whites, milk, liqueur, and vanilla extract. Pour about one-quarter of this batter evenly over the bottom of the baking dish and smooth the layer with a spatula. Bake for 8 to 10 minutes or until set. Remove the dish from the oven and arrange the berries evenly over this layer. Pour the remaining batter over the berries. Continue baking for 40 to 60 minutes or until the clafouti is firmly set and golden brown. Dust with confectioners' sugar and serve while still warm.

BERRY FOOL

Fools, dating back to the 1500s, are a simple conflation of fruit and cream or whipped cream. (If made with strawberries, the dessert is also known as Strawberries Romanoff; see page 146.) The fruit can be stewed, then folded into the whipped cream. The version given here is easier still, using juice from fresh fruit. Try whatever is at hand—raspberries, black currants, strawberries, or a blend. The traditional favorite was gooseberries, picked when not quite ripe for a sharper flavor. This version uses boysenberries, which turn the dessert a lovely magenta. The texture is something like that of ice cream on its way to melting, but lighter.

Serves six.

1 cup boysenberries
zest of 1 lemon
2 tablespoons Cointreau or Grand Marnier
6 tablespoons confectioners' sugar
1 cup heavy whipping cream

Place the berries in a saucepan containing ½ inch of water. Simmer the berries for 5 minutes, crushing them with a potato masher as they soften. Strain through a sieve, pressing with the back of a large spoon to express as much juice and pulp as possible. Stir in the lemon zest, liqueur, and sugar, making sure the sugar is dissolved. Whip the cream until fairly stiff. Stir the berry mixture into the whipped cream and chill. Serve the fool in individual dessert dishes.

BERRY GRUNT

A grunt is a type of cobbler in which small spoon-formed dumplings cook while sitting atop a simmering berry mixture—you don't need the oven. Grunts often are made with blueberries, as befits their origin somewhere between Maine and Canada's Maritime provinces; this version uses raspberries. The dessert is at its best served while still warm from the pan because the dumplings soon turn tough and less appealing.

Serves six.

For the berry mixture:

3 cups fresh or frozen berries
½ cup water
½ cup sugar
1 tablespoon fresh lemon juice
zest of 1 orange

For the batter:

1 cup unbleached white flour
2 tablespoons sugar
½ teaspoon ground cinnamon
½ teaspoon salt
1½ teaspoons baking powder
3 tablespoons chilled butter
¾ cup milk
1 cup heavy cream, whipped (optional)

You'll need a large saucepan—10 inches across will do—with a close-fitting lid. If you aren't so equipped, you can use a broad skillet and improvise a top by folding a sheet of aluminum foil snugly around its perimeter; use hotpads or oven mitts to do this safely. Or take a tip that first appeared in an old *Joy of Cooking* and use an ovenproof glass pie plate for a lid. It will contain the heat while allowing you to watch the progress within.

Simmer the ingredients for the berry mixture in the saucepan over low heat for 5 minutes. Cut in the cold butter with a fork until you reach a mealy consistency. Stir in the milk just enough to make a stiff batter. Drop the batter in six spoonfuls onto the simmering berry mixture to form the dumplings, allowing some space between them to expand. They will take on a shaggy form, and there's no need to fuss over their shape. Put on the lid and continue to simmer for another 15 minutes without peeking. Depending on how tightly the lid fits, you may or may not hear the grunt of escaping steam that earned this recipe its odd name. Check to make sure the dumplings are done by inserting a toothpick and seeing if it comes out clean. The raspberries will have cooked down into a rich jam. Serve warm, and top each portion with whipped cream if you wish.

BERRY COBBLER

You won't have the fun of calling your dessert a "grunt," but you can bake the ingredients from the previous recipe as a conventional cobbler. Preheat the oven to 350 degrees F. Pour the berry mixture into an oiled 9-inch baking dish and drop spoonfuls of batter onto the surface. Bake for 45 minutes or until the dumplings begin to turn golden brown.

MOUSSE NAPOLEONS

Try this dessert when you want to present something that looks labor-intensive. Puff pastry, in case you haven't made its acquaintance, is a multi-layered sheet of dough that expands in the oven, sold in frozen form.

Serves six.

1 cup low-fat ricotta
1 teaspoon orange extract
½ cup heavy cream, whipped
¼ cup blackberry jam or blackberry syrup
1 frozen puff pastry sheet, thawed
⅓ cup semisweet chocolate chips
1 tablespoon butter
1 to 2 tablespoons water
confectioners' sugar for dusting

Thaw one pastry sheet, allowing about 30 minutes. Preheat the oven to 375 degrees F. In a food processor, whip the ricotta and the jam or syrup until very smooth. Add the orange extract to the heavy cream and whip until the cream forms stiff peaks. Gently fold the whipped cream into the ricotta. Refrigerate this mixture for at least 30 minutes.

Unfold the pastry sheet and use a knife to cut along the folds to make three strips. Cut each strip in half, making six rectangles in all. Place the rectangles on a baking sheet and bake for 15 minutes or until golden brown. Allow the rectangles to cool on a rack, then split them into two layers to make six tops and six bottoms. Spread the ricotta mixture on the bottoms. Put the tops in place.

Melt the chocolate, along with the butter, in a double boiler. Stir in just enough water to arrive at a consistency that will allow you to pour the melted chocolate. Drizzle the tops with it and dust with confectioners' sugar through a sieve. Serve the same day.

LEMON BALM BERRY TART

Much as the juice and grated zest of lemons figure in many berry recipes, lemon balm leaves accent the tarts made by Valery Hawk Hoffman of Sunrise Herb Remedies in Connecticut (www.sunriseherbfarm.com). If you don't have lemon balm and are familiar with another lemony culinary herb—verbena, thyme, or scented-leaf geranium—give it a try.

Serves six to eight.

1⅔ cups unbleached white flour
½ teaspoon salt
½ teaspoon baking powder
½ cup unsalted butter
2 teaspoons ice water
4 cups berries, fresh or frozen
1 cup confectioners' sugar
1 cup unsalted butter
2 eggs
1 teaspoon vanilla extract
½ teaspoon finely chopped lemon balm leaves

In a food processor, combine the flour, salt, baking powder, and butter until the mixture is crumblike. Add the water and continue processing until the mixture forms a ball. Wrap the ball of dough in plastic and refrigerate for 30 minutes.

Preheat the oven to 375 degrees F. Roll out the dough on a lightly floured surface. Press it into a 10-inch tart pan with a removable bottom, running the dough up to form sides that will contain the filling. To prevent the edges from burning, you can fold strips of aluminum foil over them. Keep the dough from lifting while baking by pricking the base here and there with a fork; you also can place pie weights, dried beans, or clean stones on the bottom. Bake for 15 minutes, take the crust from the oven, and remove the foil. Reduce the temperature to 325.

Arrange the berries in the tart shell. (If you are using thawed berries that have created a lot of juice, drain most of the juice to ensure that the tart will set.) In a mixing bowl, whisk the confectioners' sugar and butter until blended well. Beat in the eggs, then add the vanilla and lemon balm. Pour the mixture over the berries and bake for 25 to 40 minutes or until set. Allow the tart to cool somewhat before serving.

BERRY BUCKLE

A *buckle* typically is a baked dessert with the berries stirred into the batter. It also may be known as a *crumple,* both names describing the uneven surface that can form while baking.

Serves eight.

¾ cup sugar
3 tablespoons butter
½ cup low-fat sour cream
zest of 1 orange
1 egg
½ cup unbleached flour
¼ cup yellow cornmeal
¼ teaspoon baking soda
⅛ teaspoon salt
¼ teaspoon ground nutmeg
¼ teaspoon ground ginger
3 egg whites
2 cups berries
vegetable oil to coat pan

Preheat the oven to 375 degrees F. Combine ½ cup of the sugar and the butter in a large bowl; beat with a mixer until fluffy. Add the sour cream, orange zest, and egg and beat until well blended. Combine the flour, cornmeal, baking soda, and salt; add this to the sour cream mixture and beat only until the dry ingredients are moistened. Beat the egg whites and gradually add the remaining ¼ cup of sugar until peaks form. Fold the egg whites into the sour cream mixture, followed by the berries. Coat a 9-inch cake pan with vegetable oil. Pour the batter into the pan and bake for 40 minutes. The cake surface should spring back when dented. Allow to cool on a cake rack before serving.

BERRY VINEGAR

Berries can soften the edge of vinegar as well as add flavor and brilliant color. That's not to say adding a handful of berries will salvage the harsh, chemical taste of cheap vinegar. Start with a good grade of white wine vinegar.

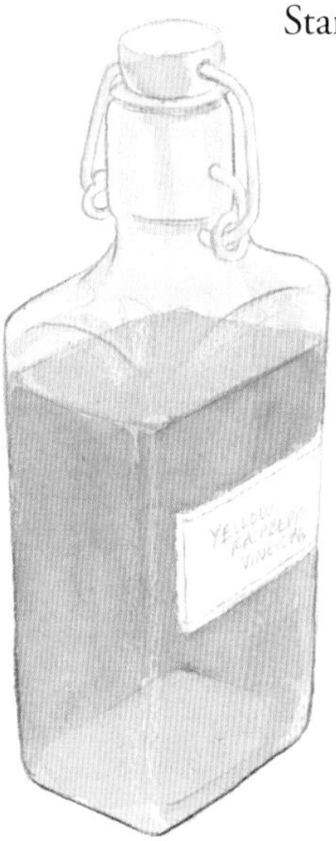

Makes about 5 cups.

2 cups any berries
4 cups white wine vinegar

Place the berries in a saucepan and crush them with a potato masher. (For harder berries, such as currants and blueberries, make a puree in a food processor instead.) Pour the vinegar over the berries. Cover the pan and allow the vinegar to steep for several days, giving an occasional stir. At the end of that period, simmer the vinegar for 5 minutes, then strain through cheesecloth. When the vinegar is cool, pour it into a handsome bottle. Store the bottle out of direct sunlight or the lovely liquid may fade to a drab color.

BERRY TURKISH PASTE

The goodness of berries can be captured in the sugary gel of Turkish paste. Here, blackberry syrup is combined with a nudge of grated ginger. Feel free to use any concentrated berry syrup, jelly, or jam. To guarantee good results, it helps to have a candy thermometer on hand. If not, you'll have to perform a bit of countertop lab work to determine that the sugar mixture has had enough cooking.

Makes about 70 cubes.

4 cups sugar
4 cups water
1¼ cups cornstarch
1 teaspoon cream of tartar
1 tablespoon fresh lemon juice
½ cup blackberry syrup
1 tablespoon finely grated ginger
1 cup confectioners' sugar
vegetable oil to coat pan

Pour 1½ cups of the water into a saucepan and stir in the sugar until dissolved. From this point, simmer the sugar mixture without stirring. Its temperature will rise as the concentration of sugar increases and water is driven off; for the paste to gel properly, the mixture should reach 240 degrees F. If you don't have a candy thermometer, you can test for that temperature by determining when the mixture reaches the *soft ball stage.* Chill a measuring cup of cool water by adding a couple of ice cubes. Drop ½ teaspoon of the mixture into the cup for each test. At first the mixture dissipates into the water; next, it turns into threadlike strands; and moments later you will see it form a ball of sugar. Take out the ball and hold it in your fingers. It should flatten without squeezing. (A hard ball indicates the sugar solution has become too concentrated.) Immediately take the mixture from the heat.

In another saucepan, thoroughly combine 1 cup of the cornstarch and the cream of tartar. Gradually stir in the remaining 2½ cups of water, taking extra care to avoid forming lumps or you'll end up with solidified cornstarch at the bottom of the pan. As you continue stirring over medium-low heat, add the blackberry syrup and ginger. Bring this paste to a simmering boil. Add the warm sugar mixture. Simmer the paste for 1 hour over low heat, stirring to keep the sugar at the bottom from caramelizing. Remove the paste from the heat.

Prepare a 9-inch square pan by lining it with plastic wrap and going over the wrap with vegetable oil. Spoon the warm (not hot) paste into the pan to form an even layer. Cover loosely with a piece of waxed paper, and allow the paste to stand for 24 hours.

On a mixing board or clean countertop, sift a mixture of the remaining ¼ cup of cornstarch and the confectioners' sugar. Lift the paste out of the pan and place it on this surface. Cut the paste into cubes with a knife. This will go more smoothly if you keep the blade clean and spread a bit of oil on it. The traditional way to keep the cubes from adhering to each other and everything else is to coat them with the powder of sugar and cornstarch. But if the cubes are weeping excess moisture and creating sticky little pools, the powder will turn to goo. Allow the cubes to stand out for another day or more, until dry to the touch.

Tumble the cubes in the powder so that all six faces have a thin, nonsticky coating. Place the cubes in a cookie tin of the appropriate size, with a square of waxed paper and a dusting of powder between each layer.

BERRY ICE CREAM

Depending on how rich you like your ice cream, you can substitute light cream for the milk in this recipe, or use it instead of the heavy cream. Our family likes to add lemon zest and fresh ginger juice to berry ice cream; to make the juice, place grated ginger in a scrupulously clean garlic press and give a good hard squeeze.

Makes eight to ten servings.

2 cups berries
1½ cups milk
1½ cups heavy cream
½ teaspoon vanilla extract
½ teaspoon fresh ginger juice
zest of 1 lemon
¾ cup sugar

Follow the instructions that come with a frozen dessert machine.

BERRY LIQUEUR GRANITA

This dessert satisfies in the manner of a sorbet or ice cream but spares you the need to plug in and clean a frozen dessert maker, if in fact you have one. The light texture is created simply by scratching at a block of frozen fruit syrup just before serving. If you double the recipe, you can keep enough granita on hand to prepare for more than one meal. Return the unused portion of the block to the freezer. Raspberries seem especially suited to a granita, but use any berry you'd like.

Serves six.

2 cups water
2½ cups fresh or frozen red or black raspberries
½ cup sugar
2 tablespoons Cointreau or Grand Marnier
zest of 1 lemon

In a saucepan, bring 1 cup of the water to a boil and simmer the berries for 5 minutes to help release their juice. If you wish, remove the seeds by pouring the berries and liquid into a strainer. Return the juice to the saucepan and bring it to a simmering boil. Add the remaining cup of water and the lemon zest, and stir in the sugar until dissolved. Allow this syrup to cool, stir in the liqueur, and place it in a covered container in the freezer. After an hour or so, stir the partially frozen granita to minimize separation, then freeze it solid. (You may have to turn down the thermostat of your freezer to do so.) Prepare the granita for serving by vigorously going over it with a fork to create a light, fluffy texture. If the block is frozen so hard that you have difficulty fluffing it up, allow it to stand out to soften somewhat.

BERRY SPRITZER

This is a lighter, more healthful variation on soda pop, with more substance than a flavored seltzer. Pour 2 tablespoons of berry syrup in a tall glass. Add two fingers of chilled plain seltzer and give a stir. Add ice cubes and fill with seltzer.

BERRY VODKA AND LIQUEUR

For centuries before mechanical refrigeration, distilled spirits provided a way to preserve the flavor (if not the nutritional value) of berries. Alcohol has an antimicrobial effect that helps to prevent spoilage.

Half fill a glass jar with fresh or frozen berries, add unflavored vodka nearly to the brim, and put on the lid. Allow the jar to sit for at least a couple of weeks of *maceration,* as this stage is known, in a cool, dark place. Give the jar an occasional shake if you think of it. Pour the liquid through a coffee filter or a strainer lined with clean cheesecloth. If the fruit taste is too strong for your liking, you can either add more vodka or stir in a tablespoon or two of sugar. Then bottle the vodka and allow it to age further. With time, the fresh and fruity taste gives way to subtle, burnished flavors; by sampling from month to month, you can decide on the optimal aging period.

To make a berry liqueur, sweeten the flavored vodka with *simple syrup*. In a saucepan, stir sugar into an equal amount of water over low heat until the sugar dissolves completely. Allow the syrup to cool and add it in small increments to the vodka until you reach the level of sweetness that suits you. Refrigerate the unused syrup; it figures in many cocktail recipes and also can be stirred into tart berry-based spritzers. Although the liqueur is ready to be enjoyed right away, it may benefit from maturing for a few months in a well-sealed bottle. See specific suggestions on making berry-flavored vodkas in the chapters on black currants, elderberries, juniper, and strawberries.

TUTTI FRUTTI LIQUEUR

One of the easiest means of preserving berries is to drop them into a crock of sweetened brandy. The alcohol and the added sugar preserve the berries, which gradually give up their essences to the liquid. Begin by pouring a bottle of unflavored French brandy into a glazed crock or, less picturesque, into anything from a stainless steel soup pot to my choice, a wide-mouthed plastic picnic jug. Then add any berries you have on hand (the name of the recipe means "all fruits" in Italian), along with an equal volume of sugar. You can cut up or crush the berries to speed up this process of maceration if you wish. Give a good stir to help dissolve the sugar. Alternately, you can keep a jar of simple syrup in the refrigerator as the sweetener; see the previous recipe. Each time you add berries or other fruits, top off with brandy as needed to keep these ingredients covered. To help keep the fruits submerged, place a weighted saucer on them. Store the tutti frutti vessel in a cool, dark place, but handy enough so that you can give it an occasional stir.

Some tutti frutti recipes insist that you hold off sampling the goods for months, but even after just three or four weeks you'll be rewarded with a brilliantly colored, full-flavored liqueur. You can keep adding berries, brandy, and sugar indefinitely, pouring off enough occasionally to bottle for long-term storage or gift giving.

BLACKBERRY WITH THORNS

CHAPTER 2

Blackberries

The wild blackberry truly is one wild berry. With its aggressive thorns, sprawling habit, and complex flavor, it has presence—both in the landscape and when eaten fresh or in recipes. To the Iroquois, the juice from the rugged plant was believed to make a person resistant to cold weather. They have a legend in which a young boy chased away Hatho, the frost spirit, by throwing hot blackberry sauce in his face; ever since, the spirit has remained hidden in his northern lair from when the blackberries blossom until after the fruit is fully ripe.

Thoreau, our savant of meadow and forest, named this fruit the favorite of his berries, placing it in his personal pantheon along with the white pine and the hermit thrush. But it would be like Thoreau to pick a somewhat difficult berry, just as he claimed to prefer pocked, scabby wild apples to cultivated orchard fruit.

Until not all that long ago in human history, blackberry bushes were regarded as an aggressive weed, not to be invited into the garden. Attitudes became more charitable as open land began to disappear in America's populous areas. People began to dig up blackberry plants to place in a corner of the yard where they couldn't get into much trouble. Pampered with good soil, the briars tend to be more productive. Still, these berries weren't considered on a par with other species. In *The Anatomy of Dessert* (1934), Edward A. Bunyard bashed the blackberries of his native Great Britain, calling them a "fruit for out-of-doors" rather than something suitable to serve as dessert, while the American varieties "are good for cooking purposes only." It may be that the blackberry's well-known reputation as a home remedy for digestive complaints (diarrhea in particular) made it difficult for people to think of it as a delicacy. Up until the late 1800s, blackberry wine and brandy were commonly sold for relief from digestive complaints. If these products or the fresh berries couldn't be had, both city and country folk would brew a digestive tea from blackberry roots. An over-the-counter tincture still is commercially available.

BLACKBERRIES IN THE YARD

The best blackberries you'll ever taste may be lurking in a hedgerow near you. The cultivated, named varieties aren't necessarily any more flavorful than those that have been growing wild for centuries, and some may be noticeably less so. If you find a productive, good-tasting patch and can get permission from the landowner, you might try transplanting a few blackberries to your garden.

If you want to grow blackberries, wild or cultivated, keep in mind that these plants will sucker freely, requiring you to cut, mow, or pull the new canes. Relative to raspberries, blackberries can take more heat and will put up with less cold, making them a better choice for backyard growers in warmer zones. Prepare the ground and set out plants as described for raspberries, Chapter 11.

BLACKBERRY FALL FOLIAGE

Blackberries grow and spread in two different ways. Those varieties with an *erect* aspect send up their new canes from the roots. *Trailing* varieties have relaxed canes that sprout if allowed to arch and reach the soil; these are also known as dewberries, and loganberry and boysenberry are of this type. Trailing blackberries thrive particularly well in the northwestern United States.

If you've chosen erect blackberries, allow them to grow unpruned their first year after planting. In the following years, pinch them back to a length of three or four feet. Prune away the bearing canes at the soil line after you've picked the summer's crop. When winter temperatures begin to moderate, take out any canes that have strayed outside their rows, and thin canes as necessary so that there are no more than a half dozen per foot of row. Trim back side branches to twelve to eighteen inches.

Trailing blackberries need the support of a trellis. Just a single strand of wire between posts will do. Tie the canes to the wire. After the harvest, your next maintenance task depends on your climate. In warmer zones, prune away the canes that have just borne fruit, then tie up the new canes in their place. In the north, you can protect these

CHESTER BLACKBERRY

tender new canes from freezing temperatures by allowing them to lie on the ground, covering with mulch, and taking advantage of the insulating effect of snowfall. Tie them back up in spring.

You can grow thornless blackberries that produce mammoth fruit, Chester being a well-known variety. So why would anyone bother with snarly old blackberries? It seems that when it comes to berries, you can't have it all. What a variety gains in size or yield or hardiness it may lose, if just a little, in flavor or complexity—qualities that are hard to quantify and impossible to show in catalog illustrations. The Chester blackberries on our place produce lavishly and dependably, year after year, and they even offer large pink blossoms in the bargain, but the berries just don't have the inviting flavor of less-enthusiastic bearers. Some summers, if our other berries and the hedgerows are productive, the Chesters go unpicked.

BOYSENBERRIES

The boysenberry is an exceptional blackberry-raspberry hybrid, discovered in 1920. It was languishing in an abandoned California field that had been tilled by farmer Rudolf Boysen. You can see and taste a bit of both parents in this offspring, but the boysenberry is its own berry, so to speak. The color is a soft reddish purple, and the taste is fruity and even a bit winelike.

MARIONBERRIES

The marionberry was developed in Oregon through the efforts of the USDA and is named for an Oregon county. From its introduction in the 1950s, it has become the most popular named blackberry. Production still is centered in Oregon, where the berry's pronounced, distinctive flavor finds its way to jams, syrups, and many other products.

LOGANBERRIES

Another gifted cross thought to be between blackberries and raspberries is the loganberry. In 1881, a California judge named J. H. Logan spied the unusual canes growing in his home garden. The dark berries have a pronounced flavor and they are used primarily for

BOYSENBERRY

juice, pies, and wine. In parts of the West Coast, loganberries have been supplanted by another hybrid, the youngberry, developed in 1905 by B. M. Young of Morgan City, Louisiana. To travel a little further along this complex family tree, a variety called Black Logan was crossed in 1950 with the youngberry to create the large and slender olallie in 1950. The olallie in turn was crossbred to create the marionberry.

BLACKBERRIES IN THE KITCHEN

When picking blackberries, give them a taste test before you start dropping them into the berry bucket. They can look quite ripe but still be tart and insipid. Allow them to mature fully to bring out their flavor at its most intricate and intriguing.

Fresh blackberries are not readily available in stores because their quality suffers during shipping. Commercial varieties are sturdier than the wild form, but nevertheless they can turn soft, mushy, and moldy soon after being picked. When shopping for blackberries, be sure to check the bottom of the container to ensure that there are no moldy or crushed berries. Try to use them the same day that they are gathered or purchased, or by the following day at the latest. If that won't be possible, spread them out on a cookie sheet, freeze them, then store them in the freezer in plastic bags. And as with any berry, you can stop time (so to speak) by preserving their goodness as a syrup or jelly.

BLACKBERRY SOUP (MAKVLIS SUPI)

This Polish soup hits to all corners of the palate for a well-balanced flavor that its odd roster of ingredients wouldn't suggest. Poland is active in developing blackberries with improved flavor and productivity, and foraging for the wild berries in the countryside remains a popular activity.

Serves six.

2 tablespoons olive oil
1 medium onion, chopped finely
2 cloves garlic, minced
1 tablespoon balsamic vinegar

3 tablespoons ale (or beer or red wine)
3 cups blackberries, fresh or frozen
1¼ cups water
1 medium cucumber, peeled, seeded, and diced
2 tablespoons mint, minced
1 cup cilantro, finely chopped
½ teaspoon thyme, finely chopped
½ teaspoon salt
sour cream or yogurt as garnish

LOGANBERRY

In a skillet, sauté the onion in the oil until clear. Add the garlic and sauté for another minute. Add the vinegar and ale and stir to deglaze the skillet, then place the mixture in a bowl.

Simmer the berries in ½ cup of water for 5 minutes, or longer for frozen fruit. Pour through a sieve and press out as much juice and pulp as you can with the back of a large spoon. Add this liquid and the water to the bowl. Stir in the cucumber, mint, cilantro, thyme, and salt. Refrigerate for at least 6 hours. Serve cool with sour cream or yogurt.

BLACKBERRY MARTINI

There are cuter and more elaborate ways to work up a blackberry martini. But this version is sturdy enough to become a nightly standard. There is just enough blackberry to float a perfume over the sterner scent of gin. The color is indescribably lovely. These ingredients are knitted together by Lillet, the French aperitif.

Make your own blackberry syrup or buy prepared syrup. Alternately, dissolve blackberry jelly in an equal amount of warm water.

3 ounces gin
1 tablespoon Lillet
2 teaspoons blackberry syrup
lemon twist

Chill a cocktail glass. In a cocktail shaker half full of ice cubes, shake the ingredients for 20 seconds or so. Pour into the glass and garnish with the lemon twist.

BLACKBERRY TONIC

Blackberries have a quieting effect on the lower digestive tract. If you can't find a commercial over-the-counter blackberry tonic, it is easy (and flavorful) to come up with your own. The one unusual ingredient in this recipe is zatar, a spice blend popular in the Middle East for topping bread as well as easing stomach complaints. Zatar formulae vary from one country to another, but the principal herb tastes something like oregano. Look for zatar at Middle Eastern specialty stores or purchase it through mail order.

Makes about three cups.

8 cups fresh or frozen blackberries
3 cups water
2 cups sugar
1 tablespoon Angostura bitters
2 tablespoons zatar
1 teaspoon ground ginger
1 teaspoon ground allspice
1 teaspoon ground mace
1 teaspoon ground cloves
1 teaspoon ground cinnamon

In a saucepan, simmer the berries in the water for 5 minutes, and crush them with a potato masher as they soften. Pour the berry juice through a strainer. Return the juice to the saucepan, add the sugar, and simmer while stirring until the sugar is dissolved. Add the remaining ingredients, cover the pan, and simmer for 15 minutes. Pour the tonic through cheesecloth and bottle it. For relief, take 1 tablespoon of tonic or stir 2 tablespoons into a small glass of seltzer water. But don't pigeonhole it as a nostrum. The tonic also can be used as a marinade, condiment, ice-cream topping, or flavoring for mixed drinks of your own invention.

BLACK CURRANT

CHAPTER 3

Black Currants

The black currant inspires a love-hate reaction in people, putting it in the good company of garlic, gin, licorice, and certain challenging cheeses. The flavor has been described as "off-putting," "foxy," "repulsive," "flamboyant," "peculiar," "disagreeable," "mawkish," and perhaps putting too fine a point on it, redolent of cat urine. It has been said that while black currants may become more palatable with cooking, the same is true of old shoes. Fruit expert U. P. Hedrick, who wrote the profiles of hundreds of berry varieties, was unimpressed by this one. "Few Americans have tasted the black currant," he wrote in 1925, "and few of those who have would care for a second taste." In *The Fragrant Path,* published several years later, respected gardening author Louise Beebe Wilder placed both American and European black currants on her list of plants with an "evil odor." More recently, Vermont nurseryman Lewis Hill dismissed them altogether in his book *Fruits and Berries for the Home Garden.* "I'd avoid these," he wrote, and he promptly went on to discuss other berries.

And yet to some of us, the sweet-sour flavor and curious scent are ambrosial, with complex overtones of pine or cedar, sandalwood, black cherry, and mince pie. The tiny bell-shaped blossoms are an ingredient of Chanel No. 5. And in a recent *New York Times* article on an up-and-coming Manhattan street, the reporter mentioned the scent of cassis wafting through the air as evidence of gentrification. So this dark berry may be on its way to redemption.

It is true that somewhere in the mix of flavors, like the unsettling eleventh note in a modern jazz chord, there floats a meaty muskiness that just doesn't seem berrylike. But our family has learned to love black currants, which we found growing on our property when we moved in. We find them the easiest form of fruit to grow and harvest, without exception. Our dwarf apple trees are pest magnets, the cherry trees were overtaken with a horrid black gum, the raspberries and blackberries run rampant, and we found strawberries to be too much bother. But our black currants can be counted on to come through every June with only an occasional, casual pruning.

Most black currant varieties come from northern Europe, where the berries have been a vital source of vitamin C. Russia, Poland, and Germany remain the leading producers of black currants. The French use black currants to produce crème de cassis, a liqueur with a grown-up taste missing in many other fruit brandies.

The British drink their black currants in the form of a sweetened nonalcoholic syrup marketed under the trade name Ribena. It can be diluted to make a refreshing drink with a good measure of vitamin C, or poured straight over ice cream, cereal, and pancakes. Originally developed for use in hospitals and maternity wards, Ribena was regarded as a "wartime tonic" in World War II when citrus imports became scarce. Because of Britain's historical influence in Asia, your best chance of finding Ribena may be in a local East Asian food market. British kids continue to drink a lot of Ribena, enough so that bottles bear a warning that it shouldn't be given to the young who still are using a "dummy"—not of the ventriloquist's sort, but a pacifier. The acid in Ribena may harm young teeth, and a low-acid variety of Ribena is now marketed for this reason. The British also enjoy black tea scented with the berry and black currant preserves. The oddest use I've seen for black currants is as the flavoring for enormous candy slugs sold in a shop in Yorkshire. The things were supple and black, and remarkably similar to the ones you'd see crawling along the local hiking trails.

Not all of Europe is in love with the black currant. The farther south you go, the less their appeal, presumably because citrus fruits become more available. According to the Italian authors of *The Complete Book of Fruits and Vegetables,* black currants "are not particularly good"; and as if to support that opinion they added that the berries once were thought to spawn stomach worms.

Black currants never had been very popular in the United States, and then production plummeted in the early 1900s with concern over white pine blister rust, a disease hosted by currants and afflicting white pine trees. With the development of resistant varieties, the federal government lifted its restrictions on selling the plants. But some states still have regulations in place, as you'll see if you page through a mail-order nursery catalog. Here and there around the country, amateur and commercial growers are trying to stimulate the nation's appetite for black currants. Ed Mashburn, who grows a number of varieties in a Pennsylvania plot, finds a range of flavors nearly as broad as that offered by apples and grapes. He predicts that American supermarkets may one day stock black currant candy, yogurt, ice cream, and cough drops. Among his favorite varieties are Ben Lomond (Ribena, the sweetened black currant concentrate, is made from this one) and Minaj Smierou, a Bulgarian introduction. (See Chapter 12, Red Currants, to read about this relative.)

NATIVE CURRANTS

Among the plants that Lewis and Clark brought back from their trip to the Pacific Coast were the golden or buffalo currant *(Ribes aureum)* and the clove currant *(R. odoratum).* The golden yellow trumpet-shaped flowers are their chief talent. Their berries are fruitier and milder than those of European black currants, and they are better enjoyed in recipes than eaten fresh. Thomas Jefferson was delighted with the discovery of these bushes and he grew them at Monticello. Native black native currants eventually became a popular foundation planting because breezes would bring the clovelike fragrance of the blossoms in through open windows. (With this in mind, we've planted a clove currant shrub just outside a kitchen window, along with fragrant moonflowers, clematis, and an old-fashioned rose; the pleasant scents do enter the home, if subtly and for a short period of the year.) A variety of *R. odoratum,* Crandall, selected back in 1888 for its large berries, is still commercially available. Unlike another early-blooming shrub with yellow flowers, forsythia, these currants won't spread aggressively and become a nuisance.

THE MILD-MANNERED JOSTABERRY

If you find the flavor of the black currant to be too assertive, give jostaberries a try. This recently developed hybrid is a cross between a black currant and a gooseberry, wedding their flavors while sparing you the intensity of the former and the thorns of the latter. Our jostaberries grow vigorously and are highly productive without fail, so that we always have some stashed in the freezer—and they hold up well when frozen. Set out two or more plants to ensure pollination.

BLACK CURRANTS IN THE YARD

Black currants are best planted in the fall. They can be arranged as a hedge, spacing the shrubs every three to five feet. By growing a mix of currants and gooseberries, you'll have the luxury of a long harvest period. First to ripen are black currants, then red and white currants, and finally gooseberries.

Prune the plants in spring by cutting back branches to maintain the shrubs' form and leaving only a half dozen or so shoots to grow from the plant's base. You'll note that the berries appear on both the pale new shoots and the darker two-year-old branches. Remove older branches that are no longer producing. You can use a shovel to chop off unwanted canes that have developed roots and grow new plants from them. I set out a hedge along a stone wall by using these free plants, and within a couple of years they were well established and productive. Currants benefit from annual applications of manure or a complete fertilizer, along with a layer of mulch.

Currants grow on *strigs,* a word that looks like a typographical error but describes the thin stem from which the berries hang. To make harvesting a little easier, pull or snip off the strigs, then bring them indoors to remove the individual berries. You can prolong the harvest by draping fabric dropcloths over the bushes; ripening takes longer in the shade.

BLACK CURRANTS IN THE KITCHEN

Black currants become milder with cooking because the heat reduces the acidity, contributing to a better sweet-sour balance. Don't overlook the leaves, if you have access to the plants; try using them sparingly in any black currant dish or beverage. They have a strong scent that's similar to the berries but with their own fresh, green character.

We keep black currants on hand, fresh and frozen, to toss into any dish that might benefit from a kick of exotic flavor and brilliant color. For example, a handful of them works well with butternut squash in the sweet-sour Kashmiri pilau that's become a favorite recipe of ours. We also use black currants in oatmeal, muffins, homemade ice cream, and a vodka-based liqueur.

Ounce for ounce, black currants have from three to five times the vitamin C of

JOSTABERRY

oranges—as you might expect, given the intense flavor of the berries and their relatively small size. Black currants also rank near blueberries in their level of antioxidants. Long before nutritional studies established just what made the berries healthful, black currant jelly was used as a remedy for sore throats. A drink of jelly or jam stirred in water was considered to have a cooling effect on fevers.

By the way, although black currants look as though they might shrivel up into the dried currants that come in boxes at the supermarket, there is no relation between the two. Dried currants are in fact grapes. They got their misleading name in a roundabout way, having been associated long ago with the Greek city of Corinth. In a centuries-long version of the parlor game Rumor Down the Alley, *Corinth* became *currant.*

BLACK CURRANT CORN PANCAKES

Bored with standard-issue pancakes? Try this summertime recipe with its crunch of fresh corn and tangy bursts of black currants. To enjoy these cakes year-round, we freeze corn cut from the cob, placing it in self-sealing plastic bags. It's easy to spoon out frozen corn and berries as needed for a recipe.

Serves four.

1¼ cups milk
1 whole egg or 2 egg whites
1 tablespoon sugar
½ teaspoon vanilla extract
1 tablespoon olive oil, and as needed for the skillet
1 cup unbleached white flour
1 teaspoon baking soda
½ teaspoon salt
5 heaping tablespoons fresh or frozen corn
3 heaping tablespoons fresh or frozen black currants
syrup and butter

Place a skillet over medium-low heat and spread oil on it if necessary to prevent sticking. In a mixing bowl, whisk together the milk, the 1 tablespoon of oil, the egg, and the vanilla extract. Stir in the flour until any dry lumps disappear. Fold in the corn and berries. There is no need to allow frozen ingredients to thaw beforehand if you make relatively thin cakes.

Pour batter to make conventional-sized cakes or one large, pan-sized cake that will equal an individual serving. Flip when the bottoms are golden brown, then cook the top side. Serve with syrup and butter or toppings of your choice.

BLACK CURRANT TEA

Today's commercial black currant tea blends tend to have a perfumey, one-dimensional scent. For a well rounded brew, there is a quick option—stir in a teaspoon of black currant jelly per cup of black tea—and the more involved one below.

Makes two cups.

1 rounded teaspoon loose black tea
6 black currants, mashed between two spoons
3 black currant leaves, pinched and rolled between your fingers
½ teaspoon sugar, or to taste

Put the ingredients in a teapot and add boiling water. Give the tea a stir and allow it to brew for 3 to 5 minutes. Do as the tea-loving Moroccans do and oxygenate the tea to give it life. Pour one cup, return it to the pot, then serve by pouring from as high as you dare.

STOUT AND BLACK

Not everyone cares for stout. Not everyone cares for black currants. But the minority who are devoted to both may be intrigued by this Irish-French hybrid, served by some pubs in Paris.

1½ ounces crème de cassis or black currant syrup
12 ounces stout, chilled

Pour two fingers of stout in a glass, followed by the cassis or syrup, and mix well. Add the rest of the stout.

PARISIAN COCKTAIL

Also known as a Kir Martini, this hauntingly hued straight-up drink brings the dark warmth of black currant to the somewhat clinical chill of gin.

Makes one cocktail.
3 ounces gin
1 tablespoon Lillet Blanc
1 tablespoon crème de cassis or black currant syrup
lemon twist

Chill a cocktail glass. In a cocktail shaker half full of ice cubes, stir or shake the ingredients. Pour into the glass and garnish with the lemon twist.

BALLET RUSSE COCKTAIL

This mixed drink has far more savor and richness of scent than commercial black currant vodkas.

Makes one cocktail.
2 ounces vodka
½ ounce crème de cassis
juice of quartered lime
lime twist

Prepare as for the previous Parisian Cocktail.

POMPIER HIGHBALL

In *The Stork Club Bar Book,* written in the 1940s, Lucius Beebe classified this all-but-forgotten cocktail as a "restorative beverage" without suggesting what its curative powers might be. Also known as a Vermouth Cassis, the drink was popular in French-occupied Vietnam, as recalled in Graham Greene's *The Invisible American.*

Makes one cocktail.
3 ounces dry vermouth
1 tablespoon crème de cassis
seltzer
lemon twist

Fill a glass with ice, add the crème de cassis, then the vermouth, then the seltzer. Give a gentle stir and add the twist.

CRÈME DE CASSIS

French crème de cassis has a complex flavor that sets it apart from other berry and fruit liqueurs, but that's not to say it is at all difficult to sip. Aging tames the assertiveness of the fresh berry. To make your own, half fill a glass jar with the berries, then add unflavored vodka to the top. Include a few black currant leaves as well, if you have them. Allow the berries and vodka to sit in a cool, dark place, and give the jar a shake from time to time. Have a taste after a month or so. If you like what you've made, you can stop at this stage. To make a liqueur, add simple syrup (see page 21) to taste. With its sprightly taste and good clean vegetal scent, homemade crème de cassis is markedly different from the commercial version. For a smoother variation, you might try the recipe said to have been used by Alice B. Toklas, friend of Gertrude Stein and Ernest Hemingway. She moderated the flavor by including one part raspberries to eight parts black currants.

KIR

A splash of crème de cassis adds color and zest to white wine. This drink takes its name from the one-time mayor of Dijon, France, located in an area that produces the liqueur.

Makes one cocktail.
5 ounces chilled dry white wine
1 or 2 teaspoons crème de cassis
lemon twist

Pour the wine and liqueur into a wine glass, stir, and garnish with the lemon twist.

WILD BLUEBERRY

CHAPTER 4

Blueberries

Gardening writer Louise Beebe Wilder quoted an unnamed berry lover who described the blueberry as "celestial jelly wrapped in heaven-hued filament." The flavor is complex: fruity and herbal and just sweet enough. The berries grow on shrubs that act domesticated, keeping their growth within bounds. They look attractive in the home landscape, and not as if they've been put there only to generate something to eat. The delicate bell-shaped flowers somewhat resemble those of lily of the valley. Come fall, the leaves of some varieties turn brilliant yellow, orange, or scarlet.

Blueberries and their huckleberry relatives were an important crop for Native Americans. They made bread from a mixture of cornmeal and the berries. In the Northeast, the berries were preserved by drying them on racks of sticks over smoky fires. The berries could then be used to make cornbread throughout the year. Into the 1800s, New England's wild blueberry harvest was at least as important as the apple harvest according to Henry David Thoreau. Children could be expert pickers, and it was their responsibility to secure berries for the family. Not surprisingly, they tended to sample some of the loot on the walk home, and Thoreau noted they had a trick of giving the container a shake so that the berries would take up more volume.

Blueberry pickers jealously guard their favorite overgrown fields. Just west of the Adirondacks in upstate New York, Pooniel Bumstead likes to cook with berries but can't find many of the wild blues around her place. She is grateful to a friend who drives her to a prime blueberry spot each summer—but always by a different roundabout route so that is impossible for Pooniel to know just where she is.

BLUEBERRIES, WILD AND CULTIVATED

Broadly speaking, there are two sorts of blueberry—the pea-sized wild berry, and the marble-sized cultivated berry.

The little wild berries are grown on lowbush plants native to the northeastern United States and adjacent Canada. They flourish in acidic, barely fertile soil. To keep these fields producing well, commercial growers mow or burn them in alternate years. Because lowbush plants are indeed low to the ground, they tend to be protected from cold winter temperatures by an insulating blanket of snow. Today, Maine ranks first among the states in blueberry production, supplying about half of the lowbush blueberries produced in the United States and Canada.

BLUEBERRY BLOSSOMS

The larger cultivated berries are harvested from highbush plants. In the 1940s, they began taking over much of the market from the wild harvest. Michigan is the principal source of cultivated berries.

Which is better? The larger berries tend to cost less and they are more readily available fresh. When it comes to flavor, however, wild blueberries are said to have the edge. They hold up well in baked recipes. And they recently have been found to be roughly twice as rich in certain healthful compounds, ounce for ounce.

THE NUTRACEUTICALLY GIFTED BLUEBERRY

Blueberries have been known to be little storehouses of vitamins (vitamin C in particular) and various minerals. Now a widely publicized study by the Agricultural Research Service reports that blueberries were highest in certain *nutraceuticals* among all the berries, fruits, and vegetables tested. This newly coined term describes compounds in plants that are credited with disease-fighting properties. One group of these nutraceuticals, the anthocyanins, are largely responsible for the vivid color of the blueberry.

The benefits may include reducing cancer risk and resisting the effects of aging. There even is evidence that blueberries contain nutrients associated with improvements in mental functioning, particularly the powers of concentration and memory. A researcher at the USDA Human Nutrition Research Center on Aging at Tufts University suggests that a diet consisting of 1 or 2 percent blueberries "may reverse short-term memory loss and improve motor skills." If that sounds like a trifle—not much more than the usual scattering of berries atop your morning cereal—think again. In the United States, we consume an average around six ounces of fresh blueberries per year, plus another eleven frozen, for a total of just over a pound. To put that in perspective, each of us tosses down fifty-eight pounds of french fries on average.

BLUEBERRIES IN THE YARD

Blueberry bushes are attractive enough to earn a place in the yard as landscape plants. They grow relatively slowly, taking a few years to come into maturity, and don't require a great deal of pruning. Before choosing blueberry varieties for the home, consider the mature height of the shrubs. *Low-bush* blueberries include most wild shrubs and usually grow to no more than two feet. They can serve as a ground cover. Or, planted in a row along a walk, they make a tidy edging.

Highbush blueberries have been bred over the years for their size, reaching a height of eight feet or more. They include most of the varieties sold both to homeowners and to commercial growers. Highbush varieties generally can be grown in areas with warmer winters, and so-called low-chill highbush varieties have been developed to thrive in the South. Highbush varieties can be set out as a tall hedge.

Half-high (or mid-high) blueberries are hybrids with the cold-weather tolerance of wild lowbush plants and the larger berries of highbush blueberries. They grow from two to four feet high. Half-highs put the berries where you can reach them easily—not too high, and not too low—and their moderate size suits the home landscape.

You don't need a spacious backyard to grow blueberries. Dwarf varieties can be grown in pots for a good-looking plant that also produces a modest yield of berries.

A good, chilly winter is necessary for most blueberries, but if you live in the Southeast or California, consider planting *rabbit-eyes,* with their tolerance to heat, drought, and soils with a relatively high pH. The berries got their name from the pinkish cast to the fruit. Cultivated varieties can reach twenty feet in height, making harvest a bit more awkward, but you can buy commercial varieties that have been selected for compactness.

IN ASSORTED COLORS AND FLAVORS

If you've seen one blueberry, you haven't seen them all. Some blueberries become a standout in the fall, their foliage doing what sugar maples do on a larger scale. Bluegold is a compact shrub that attracts attention each fall with its bright yellow foliage, then yellow twigs in winter. Northsky is distinguished by its red foliage.

Blueberries differ not only in their shape and color, but in their nutritional composition. Some have exceptionally high levels of antioxidants, and with that in mind, I've just ordered a couple of Rubel shrubs. This blueberry was discovered growing wild in the New Jersey Pine Barrens in 1912. Its fine flavor, along with the presumed health benefits, make up for the pixie-sized berries.

There may be a noticeable variation in flavor, too. Although as yet there are no blueberries redolent of bananas or hazelnuts, you might be interested in the range of flavors offered by highbush varieties. Berkeley is noted for its mild flavor, which some people like and others find wanting. Bluecrop is popular with homeowners for baking and freezing. Elliott has a tartness appreciated by jam makers. Jersey is an old standard for home growers, with sweet, modest-sized berries that are favored for baking. Bakers also like Olympia's spicy, aromatic flavor. Northland is particularly sweet. Rubel, the antioxidant champ, has a woodsy flavor consistent with its wild heritage.

There are different tastes among the half-highs, as well. Chippewa has an especially good flavor, along with a particularly pale blue color. Northcountry offers wild-tasting berries over an extended harvest period, followed by bright red fall foliage. Northsky's berries are small, sky blue, and have a good wild flavor that's on the mild side. Polaris ripens early and is distinguished by its aroma.

Finally, the berries ripen at different times, depending on the variety. By choosing a mix of early-, middle-, and late-bearing plants, you can harvest blueberries over a period of three months.

SOUR SOIL FOR A SWEET BERRY

Blueberries aren't apt to flourish in ordinary garden-variety dirt. The plants don't require a particularly rich soil, but unless it is low in pH—acidic, that is—the plants will languish.

In some areas, the soil comes close as it is. If you see wild blueberries and native azaleas growing (and thriving) nearby, that's a sign your soil is on the acid side. But the only sure way of knowing how your soil measures up is to conduct a test, using a kit available from garden supply stores and from gardening catalogs. The pH should be between 4.5 and 5.

To lower the pH of the prospective blueberry plot, turn the soil, then mix in dampened peat moss, as well as any composted oak leaves or pine needles you may have on hand. Annual doses of powdered elemental sulfur also will knock down the number. Ideally, you should allow a new bed to rest a year so that the acidity can permeate the soil.

If your soil comes in at a pH of 7.0 or higher, you can skip it altogether and plant blueberries in raised beds made up of peat moss, sand, and compost. Or, if big yields aren't your goal, you can grow blueberries in pots. Choose from the lowbush or half-highs, or try a dwarf variety such as Top Hat, with its bright blue berries and brilliant red fall foliage.

BLUEBERRY FALL FOLIAGE

If you've got bare-root plants, get them into the ground as soon as hard spring frosts are past. Plants in pots can be set out at any time during the growing season. Lowbush blueberries can be set out every foot or so; allow two or three feet between half-highs; space highbush plants four to five feet apart; and rabbit-eye blueberries typically will need eight feet. Place the shrubs closer if you want them to mature into a hedge. Those with an attractive form and fall color can be grown individually as specimens in the yard. To ensure better harvests, encourage cross-pollination by planting more than one variety. An exchange of pollen is necessary if those delicate white flowers are to produce fruit.

You don't have to worry about pruning right away. Just pluck off any blossoms you see to prevent premature fruiting. Once half-high and highbush varieties get established, prune away dead wood, overlapping branches, and branches that are broken or growing in a tangle. Bushes that are allowed to become straggly will tend to produce a disappointing yield of small berries. (Wild lowbush blueberries need pruning only every three years or so, but at that time you should aim to remove two-thirds of the branches.)

Assuming the surrounding soil is on the alkaline side, it may be necessary to add acidic materials from year to year in order to keep the plants doing their best. Once or twice a season, give established plants a dose of an acidic fertilizer sold for azaleas and rhododendrons. As you improve the soil, keep in mind that the diameter of the root area may be up to twice that of the visible part of the plant. Also, blueberry roots spread out shallowly, so cultivate around them carefully to avoid damaging the plant.

Don't start picking as soon as you see blue—the berries need a week or so to ripen fully after turning dark. A pinkish ring around the berry stem is an indicator that you should put off the harvest a few days. Also, you shouldn't have to tug the berries in order to pick them; when ready for the berry bucket, they will release readily from the shrub. (If you are making pies, you can include some unripe berries that haven't yet reached their full color.) When you get home, discard any berries that have become crushed.

Even if you don't get around to picking the berries, they probably won't go to waste. The fruits are eaten by familiar backyard birds, as well as by grouse and chipmunks.

BLUEBERRIES IN THE KITCHEN

If you are buying blueberries, look for the *bloom*—the delicate white layer of wild yeast found on fresh, carefully handled blueberries. (This yeast is entirely safe to eat, by the way.) Discard shriveled or mushy berries. While fresh-picked blueberries store better than most, holding up for one or two weeks in the refrigerator, store-bought fruit should be used promptly.

Blueberries may change color in recipes because acidic and basic ingredients have an effect on the inky hue of the beneficial anthocyanin compounds. Lemon juice and vinegar can lend a reddish cast to a dish; less pleasantly, baking soda tends to bring out a green tinge.

When using blueberries in the kitchen, it's up to you to decide if the more pronounced taste of the wild blueberry is important. These smaller berries often are favored for baking because high temperatures decimate the large, watery cultivated type, leaving muffins pocked with holes. Also, the jumbos tend to become messy when thawed, which makes them less attractive in fruit salads and as a garnish. On that topic, there is no need to thaw berries headed for pies and cookies.

BLACK AND BLUE SOUP

This soup brings together blueberries and black currants, along with the full-bodied sweetness of figs.

Serves four.

¾ cup figs
2 cups wild blueberries
2 cups water
1 tablespoon raspberry *or* balsamic vinegar
1 teaspoon tamari
¼ cup black currant syrup or crème de cassis
¼ teaspoon freshly ground black pepper
⅛ teaspoon powdered cloves
⅛ teaspoon powdered mace
⅛ teaspoon powdered allspice
2 lemons, juice and zest
yogurt or sour cream

Rinse the figs and blueberries. Cut the stems off the figs and halve them. Set aside enough berries for a garnish if you wish.

Place the water in a saucepan and add all the ingredients except for the yogurt or sour cream. Simmer for 15 minutes. Puree the soup in a food processor and set it aside to cool. Refrigerate for at least 4 hours before serving; allow the soup to return to room temperature if you wish. Pour into individual bowls, and swirl in the yogurt or sour cream for a decorative effect.

WILD BLUEBERRY PIE

This is my mother's recipe. She uses only the berries she picks from the wild, principally from the one-acre Adirondack island on which she now lives alone each summer. The island is carpeted with berries that thrive in the dappled shade beneath old-growth white pines—lipstick-red serviceberries and blue, bauble-bright clintonia berries, as well as low-bush blueberries. Only the blueberries are edible, and not many escape my mom as she crawls over the pine needle duff—literally crawls, since for the past two summers she has been recovering from hip joint replacements. She looks something like the disabled woman in Andrew Wyeth's *Christina's World*—compromised by her lack of mobility and yet at ease in the natural environment.

Natalie is a blueberry accountant, tallying every berry she picks. On average she needs 560 berries to a cup, and two and a half cups go into a pie, for a total of 1,400 berries. Armed with this statistic, she is able (and apt) to impress people enjoying her pies with the fact that 233 handpicked berries go into a slice, which works out to 16 (give or take a few) in each bite.

Serves six.

Crust:

½ cup safflower oil
½ cup milk or yogurt, plus 2 tablespoons to help crust brown
½ teaspoon salt
2½ cups unbleached white flour
1 tablespoon sugar

Filling:

2½ cups wild blueberries
½ cup sugar
1 tablespoon lemon juice
½ teaspoon ground nutmeg
2 tablespoons cornstarch
4 tablespoons water
2 tablespoons butter

Preheat the oven to 425 degrees F. Combine the oil, milk or yogurt, and salt in a mixing bowl. Stir in the flour to form a ball of dough, adding a little oil and milk or yogurt if necessary to keep the ball together. Separate the dough into two balls and place them in the refrigerator for 5 to 10 minutes. In another mixing bowl, combine the filling ingredients—the berries, sugar, lemon juice, and nutmeg. Dissolve the cornstarch in the 4 tablespoons of water and stir well into the other ingredients.

Roll out the chilled dough into bottom and top crusts for an 8- or 9-inch pie pan. To prevent sticking, place the dough discs between sheets of wax paper. Roll from the center to the perimeter, turning them over a few times, so that the discs will be formed evenly. Peel off the paper. Place the bottom round in the ungreased pie pan and trim off the dough along the rim. Spoon in the filling and top it with pats of the butter. Put the top in place, pinching along the perimeter. To help the crust develop a good color, brush the 2 tablespoons of milk or yogurt over the top and sprinkle the tablespoon of sugar. Make a few cuts in the top to let the steam out.

Bake initially at 425 degrees F for 10 minutes, then turn the oven down to 375 degrees and cook for another 35 minutes or until the crust shows good color.

BLUEBERRY BLINTZES

You can use any berry with this one, but blueberries seem particularly suited to blintzes and not just because of the alliteration.

Makes five or six blintzes.

2 cups blueberries
1 tablespoon lemon juice
2 tablespoons sugar
2 teaspoons cornstarch
1 tablespoon water
1⅓ cups milk
½ teaspoon orange extract
1 tablespoon vegetable oil

In a saucepan, simmer the berries with the lemon juice and sugar, stirring until the sugar is dissolved. Continue simmering for 5 minutes. Stir the cornstarch into the water and thoroughly stir it into the berry mixture. Cover the pan to keep the contents warm.

In a mixing bowl, combine the flour, baking soda, and salt. In another bowl, whisk the eggs and the milk, and stir in the orange extract and oil. Place a nonstick skillet over medium-low heat. Stir the liquid ingredients into the dry. Oil or butter the skillet if you wish, but that shouldn't be necessary if the surface is performing well. With a large spoon, pour one blintz into the skillet at a time, spreading the batter thinly over the bottom with the back of the spoon or a spatula. Cook each blintz just long enough to turn both sides a light golden brown. Pile up the blintz rounds as you go, wrapping them in aluminum foil and insulating the stack with dish towels until all are made. Fill them with the warm berry mixture and serve. Provide maple syrup and butter, sour cream, or toppings of your choice.

HUCKLEBERRIES

In one of those messy name switches that commonly plague the plant world, *huckleberry* means different things to different people. In the Northeast it has been an alternate name for the blueberry (*Vaccinium*). In the eastern half of the United States, botanists speak of the huckleberry as various hard, seedy species in the genus *Garylusaccia*. In the West, ask for huckleberries and you're apt to get a tough-skinned, distinctly different blueberry relative for which there is an enthusiastic regional following but modest commercial demand. This western *Vaccinium* is better enjoyed cooked rather than eaten out of hand. It lends its flavor to jams, syrups, and recipes for tree fruits.

So, why the confusing use of *huckleberry* to describe various berries clear across the continent? One possible explanation is that word is so much fun to say that its name migrated out of bounds. Thoreau suggested that we could clear up the confusion by using the exacting names that Native Americans had for each berry, they being the original berry experts. The Iroquois even used distinct terms for "berry season" and "berry picker."

Western huckleberries may seem tough and seedy to the uninitiated, but their fans will tell you that the blueberry of the East, for all its popularity, is lacking in personality. In the northwestern United States and British Columbia, you can find all sorts of huckleberry products for sale—pies, muffins, jam, candy, vinegar, and ice cream. If you are from outside the area and want to get an idea of what huckleberries taste like, try buying a bottle of huckleberry syrup or jam from mail-order firms.

The Native nations of the Northwest and British Columbia made good use of huckleberries. To separate out twigs and leaves, they used a simple but ingenious system: the berries were rolled down a flat slab of wood and into a container, leaving the debris behind. The natives mashed the berries and dried them into cakes. And in what sounds like high cuisine, they combined boiled berries with the spawn of red salmon. Today, huckleberry picking is a popular activity for many families in the region. Parents and kids troupe into the hills to their favorite spots and pick gallons of the fruit for cooking, juicing, and freezing.

Huckleberries don't often find their way into cultivation, and nearly all of the picking is done in the wild. These shrubs are difficult to transplant to the home garden, even if you can get permission to do so from a landowner or park authority. In areas where the huckleberries are native, nurseries may offer plants of a few species—dwarf, glove, mountain (or black), red, oval-leaf, and evergreen. This last species is valued as a landscaping plant because the attractive leaves stay on the plant year-round and turn an attractive bronze shade. Mail order is another alternative. Just don't make the mistake of ordering something that goes by the name of a "garden huckleberry" but bears the Latin name *Solanum melanocerasum.* It isn't related to the native huckleberry.

WHITE HUCKLEBERRY FUDGE

One day this winter, I happened to be looking out the window as a UPS driver stopped at the end of our snow-choked lane and tossed something into a drift. I walked up to investigate and found a plastic bag, within which I could see a box with a Montana

HUCKLEBERRY

return address. And then I remembered e-mailing Ellen Bryson, the Montana huckleberry enthusiast mentioned in Chapter 1, to lament that I'd never gotten closer to a huckleberry than the bottled syrup. She responded that I should taste huckleberry fudge—and here, nestled in the snowbank, was this exotic (to an easterner) treat. You can use dark chocolate, but the berries do look stunning in the snowlike setting of white fudge.

Makes nine 3-inch squares.

3 cups white chocolate chips
14 ounces sweetened condensed milk
⅛ teaspoon salt
1 cup chopped walnuts or pecans
1 teaspoon vanilla extract
½ cup huckleberries

In a heavy saucepan over low heat, melt the chips with the condensed milk and salt. Remove from the heat and lightly stir in the nuts, vanilla extract, and berries. Line an 8- or 9-inch-square pan with parchment paper. Spread the mixture evenly into the pan. Chill for 2 hours or until firm. Turn the fudge onto a cutting board, peel off the paper, and cut it into squares. Store in a loosely covered container at room temperature.

CHECKERBERRY

CHAPTER 5

Checkerberries and Snowberries

These shy woodland berries don't attract attention to themselves, but they offer hikers and nature lovers a refreshing wintergreen-flavored nibble. They also have the effect of sharpening our eyes for the tiny woodland plants that usually go unnoticed. Indeed, the only contact most people have with checkerberries and snowberries is to tred upon them with sturdy footwear.

The checkerberry is also known as wintergreen because it keeps its shiny, oval leaves through the winter. Another name, teaberry, notes that the leaves and berries can be brewed for tea. Lovely bell-like blossoms precede the lipstick-red berries. Nursery catalogs may describe the plants as a useful ground cover, but don't expect them to form a living carpet with anything like the enthusiasm of pachysandra or vinca.

For one hundred years, my clan has been nibbling at a small colony of checkerberry, growing on the rocky point of an Adirondack camp built by my great-grandparents. Each generation introduces the next to the chewing-gum flavor of the leaves, while adding a caution against sampling other berries that might be growing nearby. The young, pliable leaves are the most flavorful. You spit out the leaves after a few chews, by the way, before they take on a bitter edge.

The leaves once were a commercial source for oil of wintergreen, used both as a flavoring and externally as a soothing medicine for sore muscles. This compound is now prepared chemically. Curiously, the wintergreen scent is also found in the bark of sweet birch trees; so it is that teaberry gum and wintergreen candy have the same flavor as birch beer.

There is a curious bit of modern-day folklore associated with wintergreen. It is said that if you crunch a wintergreen-flavored hard candy in a dark room, with your mouth open, you can glimpse a green flash in a mirror. The explanation for this phenomenon is that ultraviolet light is emitted by shattering the candy's crystalline structure, and that wintergreen oil absorbs all but the green portion of the spectrum.

CHECKERBERRIES IN THE YARD

I've tried both transplanting checkerberries and buying them from nurseries, but the only success I've had is with the tiny shrub growing here on the windowsill of my home office. The berries turned red in late summer, and now in March the bronzed, tender new leaves are opening. To succeed with checkerberries outside, you'll need to provide an acidic, well-drained soil that's rich in humus, preferably under the dappled shade of evergreens.

CHECKERBERRIES IN THE KITCHEN

CHECKERBERRY TEA

Pick young leaves and give each one a pinch to help express more of the aromatic oil. For a brew with more body, you can add half the usual amount of green or white tea.

Makes two cups of tea.
¼ cup teaberry leaves
2 cups water
sugar to taste

Bring the water to a boil in a small saucepan, stir in the sugar, and add the leaves. Remove the tea from the heat and allow it to steep for 5 minutes.

SNOWBERRIES

Look for this low-growing relative of checkerberry in boggy areas where you find sphagnum moss below and spruce trees overhead. The berries look like tiny white eggplants, with a lovely waxy pallor, and they also have the wintergreen flavor. Thoreau enjoyed brewing a pot of snowberry tea while hiking in the wilderness. A century ago, the berries were collected in the wild and sold as a novelty in tiny baskets by high-end grocers. Today, people don't much bother with this sort of charming fancy, but the berries are appreciated by bobwhite and grouse, among other birds.

SNOWBERRY

CRANBERRY

CHAPTER 6

Cranberries and Lingonberries

Cranberries claim a place at the dinner table for only a limited time late in the year, when they serve as a palate-cleansing counterpoint to the heavy foods of the season's holidays. Although cranberries once were confined to jelly and relishes, they now find their way into a remarkable variety of beverages, teas, and baked goods.

Part of the cranberry's newfound visibility is clever marketing, but the berries also have exceptional health benefits. They contain antioxidants and are known to help prevent and clear up urinary tract infections. It's a widespread belief that the berry's acidity is responsible for fighting infections, but credit goes instead to a compound that discourages bacteria from clinging to cell walls. Here are some guidelines from the University of Illinois at Chicago's NIH Center for Botanical Dietary Supplement Research in Women's Health. For a preventive effect, take three ounces of 30 percent juice each day. To deal with an infection, increase that daily dose to twelve to thirty-two ounces. If it's not convenient to chug all that juice, cranberries concentrated in capsule form also will do the trick.

CRANBERRIES IN THE YARD

You don't need to have a bog in your backyard to grow cranberries. They can do well in ordinary garden soil, so long as it is fairly acidic and high in humus. If your soil is high in clay, amend it with sand. The plants don't even need to be doused with water—just treat them as you would other plants. It may not be easy to find plants locally. Ask a nursery to special-order plants for you, or go through mail-order firms. If the plants are happy in their setting, they may spread up to twelve inches a year to form a dense mat just two to six inches tall.

CRANBERRIES IN THE KITCHEN

The fresh berries hold up well in refrigeration, thanks to their high acidity and naturally occurring antimicrobial compounds. Nevertheless, you may have trouble finding them after the end-of-the-year holidays are past. Check the freezer section for frozen berries. You also can shop for the fresh berries online, ordering direct from growers around the country.

Dried cranberries are good eaten alone as a snack, with or without added sweetener and flavoring. They can be used to spice up any fruit recipe, adding color and their distinctive bitterness. Try them as well in granola and cookies, and even in salads.

CRANBERRY RELISH WITH ORANGE

Cranberries take on still more character when coupled with oranges, skin and all.

Makes about 3½ cups.

1 pound cranberries, fresh or frozen

1 orange

2 tablespoons Cointreau or Grand Marnier

1 cup sugar

The assertive flavors of this recipe require at least a day to commingle in the refrigerator. Unless the orange is certified organic or from your own unsprayed tree, wash it in soap and hot water to remove any pesticides or preservative coating. Cut the orange into eighths and pluck out the seeds. Put all of the ingredients in a food processor and pulse to make a chunky puree. Cover and refrigerate the relish until you are ready to serve it.

SAUTÉED CRANBERRY JAM

Sautéing takes the edge off the berries, and the onions add depth and richness. This smooth, savory condiment can accompany meat at any meal, perk up a sandwich, or fill an omelet when the breakfast blahs set in.

Makes 1 pint.

4 cups fresh or frozen unsweetened cranberries, pureed
2 tablespoons olive oil
2 onions, minced
zest and juice of 2 oranges
½ teaspoon ground black pepper
1 teaspoon salt
juice of 2 lemons
2 tablespoons balsamic vinegar
4 tablespoons maple syrup

Puree the cranberries in a food processor. In a skillet, sauté the onions in the olive oil until clear, then add the pureed cranberries, zest and juice of the oranges, black pepper, and salt. Continue to sauté until the ingredients begin to darken and caramelize on the bottom of the skillet. Deglaze the skillet with the lemon juice and vinegar and remove the skillet from the heat. Stir in the maple syrup. Refrigerate the jam.

SOUR CREAM CRANBERRY BARS

This recipe was developed by Sue Indermuehle of the Alder Lake Cranberry marsh in Wisconsin (http://www.alderlakecranberry.com).

Makes about three dozen bars.

1 cup butter, softened
1 cup packed brown sugar
2 cups one-minute oatmeal
1½ cups plus 2 tablespoons unbleached white flour
2 cups dried unsweetened cranberries
1 cup low-fat sour cream
¾ cup sugar
1 egg
1 tablespoon lemon zest
1 teaspoon vanilla extract

Preheat the oven to 350 degrees F. In a large mixing bowl, cream the butter and brown sugar. Combine the oats and 1½ cups flour and add them to the creamed mixture until blended. Set aside 1½ cups of this crumb mixture for the topping. Press the remaining mixture into an ungreased rectangular baking pan (about 9 by 13 inches). Bake for 10 to 12 minutes or until lightly browned. In a large mixing bowl, combine the cranberries, sour cream, sugar, egg, lemon zest, vanilla, and remaining flour. Spread this mixture evenly over the crust. Sprinkle with the reserved crumb mixture. Bake for 20 to 25 minutes or until lightly browned. Allow the bars to cool, then refrigerate them.

CRANBERRY ENERGY BARS

My brother Carter prepares for long-distance bicycle races in the Pyrenees and Dolomites by shaving his legs and making these energy bars. You can do the same—mix up a batch of the energy bars, that is—before going on hikes or taking road trips. To make the bars more nutritionally complete, Carter includes powdered vitamins; if you do so, be sure to monitor your daily intake of the bars, and don't allow children to munch them unsupervised.

Makes sixteen bars.

2 cups unbleached white flour
1 cup soy flour
1 cup brown sugar
½ cup chopped cashews
½ cup chopped pecans
1 cup dried cranberries
2 cups one-minute oatmeal
½ teaspoon baking soda
½ teaspoon salt
powdered multi-vitamins (optional)
2 sticks of butter or soy margerine,
plus 1 tablespoon to coat pans

½ cup molasses
applesauce, as needed (optional)
8 ounces berry jam

Preheat the oven to 350 degrees F. Grease two 9-inch-square pans with the 1 tablespoon of butter or margarine. Combine the dry ingredients in large mixing bowl. In a saucepan over low heat, melt the 2 sticks of butter or margarine and stir in the molasses. Add this mixture to the dry ingredients and stir until well blended. You can add a tablespoon or two of applesauce to help the ingredients hold together (loosely, rather than like a sticky cookie dough).

Spread two-thirds of the mixture in pans and press it flat with a fork. Bake for 20 minutes or until the edges start to brown. Remove the pans from the oven and spread the jam over the top. Pour the remaining mixture over the jam and press down very lightly with the fork. Return the pans to the oven until the jam bubbles around the edges and the top turns golden, about 20 minutes.

Remove the pans and allow them to cool. Cut 8 bars from each pan, wrap them in aluminum, and store in the refrigerator.

CRANBERRY ORANGE FLAN

While most flans have a mellow flavor, this one takes on a hint of sharpness from the cranberries. The berries are strained because their skins become colorless and unappealing if allowed to remain in the flan.

Serves six to eight depending on size of ramekins.

For the caramel:

1 cup sugar

¼ cup water

For the flan:

½ cup water

2 cups fresh or frozen cranberries

zest of 4 oranges

⅔ cup sugar

1 tablespoon Cointreau or Grand Marnier

1 teaspoon vanilla extract

5 eggs: 4 whole eggs and 1 yolk

1½ cups whipping cream

1 cup milk

Preheat the oven to 350 degrees F. To make the caramel for the bottom of the ramekins, place the 1 cup of sugar and ¼ cup water in a saucepan and stir over low heat until the sugar is dissolved. Stop stirring and continue to cook until the mixture turns a rich amber color. Pour equal amounts into each ramekin, tilting the dishes so that caramel adheres to the lower sides as well as the bottom. (To avoid burning yourself, you can wear a *clean* gardening glove on the hand holding the ramekin.)

Place the ½ cup of water in a saucepan and simmer the cranberries for 10 minutes. Crush the berries with a potato masher as they soften. Pour this sauce through a strainer and press with the back of a large spoon to express as much juice as possible. Place the strained liquid in the saucepan over low heat, and stir in the orange zest, sugar, liqueur, and vanilla extract until the sugar is dissolved. In a mixing bowl, whisk the eggs until frothy, then stir in the whipping cream and milk until well blended. Add this mixture to the berry mixture. While stirring to keep the berry solids suspended, fill the ramekins.

Place the ramekins in a baking dish. Add water to the dish until it comes at least halfway up the sides of the ramekins. Bake the flans for about 45 minutes, or until a knife inserted into a flan comes out clean. Remove the flans from the oven and allow them to cool to room temperature, then refrigerate for at least two hours. To serve them, upend the ramekins over a dessert dish; if necessary, gently insert a narrow spatula around the edges. Allow the liquid caramel to drizzle around the flans.

LINGONBERRIES

The lingonberry is a cranberry relative that has been overlooked in the United States, even though a species grows wild here in North America, where it is called lowbush or mountain cranberry. The plants you'd pick up at a nursery are a European species with berries of a somewhat larger size.

Tasting much like cranberries, lingonberries are too tart to eat right off the bush, and most of the commerical production goes into making jams, sauces, and syrups to which sweetener has been added.

The plants are low-growing, with small glossy leaves and bright red berries. They can serve as a ground cover if you meet their growing conditions, but as with cranberry

LINGONBERRY

and wintergreen, you shouldn't expect the plants to aggressively stake out large swaths of territory. Lingonberries grow slowly. Give them an acidic soil with a pH of around 5, a condition that probably will mean mixing in a good measure of peat moss. My own lingonberry planting succumbed to weeds—a likely fate unless you are vigilant. A layer of mulch will help control weeds and also help to prevent the plants from drying out.

In recipes, you can substitute lingonberries for cranberries, and as with cranberries they are an excellent way of bringing life to anemic fruit dishes.

HIGHBUSH CRANBERRIES

The native highbush cranberry, *Viburnum trilobum,* earns a place in this chapter only because of the popular name. It is in fact a viburnum, and the shrubs, reaching ten feet or more in height and width, look nothing like the low and creeping cranberry. The berries, however, are a brilliant red and have a tartness that brings cranberries to mind. The flavor will be at its best if you delay picking them until after the first hard frost. Alternately, you can pick them earlier and put them in the freezer. You may not have to worry about the birds beating you to the harvest; these aren't their favorite late-season berries, perhaps because of the relatively low content of dietary fats. Simmer, crush, and strain the berries to remove their seeds, and use them for jelly, jam, or syrup.

Several named varieties are on the market, selected for the quality of their berries, their fall color, or their compact form. We have a handsome, full-sized cranberry bush growing by our pond, where both the berries and the burnt-orange color of the turning leaves are on display each October.

HIGHBUSH CRANBERRY

ELDERBERRY

CHAPTER 7

Elderberries

There is an old-fashioned sense about elderberries, as though they might have become obsolete along with dial telephones and butter churns. They are a fruit that nearly everyone has heard of but relatively few have tasted. Although elderberries once were commonly grown in backyards, they didn't make the cut when berries became a year-round supermarket item. It's even rare to find these dark, rich-tasting berries at roadside stands. Why have elderberries faded from the scene?

For one, elderberries have the fate of being described in terms of something else—as a richer version of a blackberry, or as a blander version of a blueberry. Also, they can't be appreciated fresh, but need to be converted into jam, pies, or wine. The shrubs also have a well-deserved reputation for containing harmful compounds. In years past, kids would hollow out the soft pith from elderberry stems to make blowguns, but this charming bit of lore should be forgotten because the wood is toxic, as are the roots and leaves. As reported in a 1894 gardening magazine, *Meehan's Monthly*, five children in Tarrytown, New York, were fatally poisoned when they nibbled on elderberry roots. Recently, the federal Centers for Disease Control reported the serious but nonfatal poisoning of several people in California who drank uncooked elderberry juice prepared from the berries along with leaves and branches. Be sure to use only the fruit and blossoms of the plant. One last caution: do not pick the berries from the related red-berried elder. Consult a field guide if you are in any doubt over the distinctions between the two species.

Cynthia Donne Seip, a Pennsylvania farmer, writes of elderberries,

They are tall and round and blooming,
Like big brides of giant girth,
With a fragrance blown from Heaven,
And a root in solid earth.

The fragrance she refers to is breathed by white blossom clusters, known as elderblow among other names, that look something like those of Queen Anne's Lace. In Denmark, the shrubs once were planted around homes because the clean, spartan scent was believed to repel both insects and mischievous trolls. The flowers can be used to make fritters and an ever-so-slightly alcoholic drink, as described in the recipes.

ELDERBERRY BLOSSOM

Elderberries rate high in vitamins A and C, contain significant amounts of calcium and iron, and have been found to offer a good dose of antioxidants. This might explain why elderberries once were referred to as "the medicine chest of the people"—by either early settlers from Europe or Native Americans, depending on the account you read. Pennsylvania fruit grower Marguerite Hobert's memory is clear on having been administered straight elderberry juice—no sweetener—as a child to relieve a sore throat, recalling the taste as "lemon puckery."

ELDERBERRIES IN THE YARD

Elderberry bushes are unassuming landscape plants, attracting notice only when they put out their large clusters of white flowers in June and then clusters of dark berries. Reaching six to twelve feet tall and just as wide, they are suited for a larger property. If you have a cramped yard, you may not want to give up the space needed for elderberries, although the shrubs can be kept to a manageable size if you mow regularly around them to keep down seedlings and suckers. Elderberries do well in damp spots where other fruits might not. We have a bit of a swale that's good for nothing else, and our three elderberry bushes thrive there.

You can plant a seedling of the native elderberry (*Sambucus canadensis*) and let it go at that. But the European elderberry (*S. nigra*) has bigger berries; and by planting more than one named variety of this species you can have a range of ripening times rather than a glut of berries all at once. (There also are ornamental varieties of the same genus,

Sambucus, but they may not have edible berries.) Speaking of picking, you'll have to beat the birds to the crop. You can throw bird netting over the shrubs until the berries are fully ripened or pick them a bit prematurely.

The shrubs don't require much attention. To keep them vital and well groomed, take out both weak-looking new growth and older branches that are no longer productive.

ELDERBERRIES IN THE KITCHEN

Harvesting goes quicker if you snip off the bunches and bring them inside, then strip the berries with a fork.

As with mulberries, recipes for elderberries usually benefit from acidic ingredients such as rhubarb, cider, lemon juice, red currant juice, or vinegar. You also can add interest by mixing them with aromatic strawberries or raspberries.

If you are picking the blossoms for a recipe, keep in mind that the berry harvest will be reduced proportionately. Before using the blossoms, give them a good hard shake to encourage any pollen-loving insects to evacuate. Then visually inspect the flowers to make sure they are uninhabited.

ELDERBERRY VINEGAR

You can make elderberry vinegar from scratch, by brewing elderberry wine and then letting naturally occurring yeasts convert the wine to vinegar. For quicker results, simply place 2 cups of the berries in a stainless steel pot and add enough white wine vinegar to cover them. Simmer for 1 hour. Strain through a sieve and pour into bottles.

PONTACK SAUCE

This inky, potent condiment can be used with meats or to enliven any sort of dish. Some recipes call for vinegar instead of wine, or the addition of anchovies for a flavor closer to Worchestershire sauce.

Makes 3 cups of sauce.

1 pint red Burgundy wine
1 pint fresh elderberries
1 teaspoon salt
½ teaspoon mace
4 cloves
40 peppercorns
½ teaspoon brown sugar
1 tablespoon minced fresh ginger
3 tablespoons minced shallots

Place the elderberries in a mixing bowl. In a saucepan, bring the wine to a boil and pour it over the berries. Cover the bowl and allow it to stand overnight. Strain the liquid through a sieve, pressing the berries with the back of a large spoon to express more juice. Pour the juice into a saucepan, add the remaining ingredients, and simmer over low-medium heat for 10 minutes. Pour into a bottle and refrigerate.

PRINCE OF WALES KETCHUP

Before ketchup became codified as a mildly spiced tomato product, nearly any berry, fruit, or vegetable might find its way into a recipe. This venerable ketchup is based on elderberries.

Makes about 1½ quarts.

1 cup balsamic or berry vinegar
1 medium onion, minced
½ teaspoon hot sauce
2 tablespoons sugar
1 teaspoon salt
½ teaspoon ground black pepper
¼ teaspoon ground mace
¼ teaspoon ground nutmeg
5 cups elderberries

Simmer the minced onion in the vinegar for 5 minutes. Add the other ingredients and simmer for 15 minutes, crushing the elderberries with a potato masher to help release their juice. Remove from the heat and puree in a food processor. Refrigerate as much ketchup as you'll be able to use in the next few weeks and freeze the rest.

ELDERBLOW FRITTERS

The flowers of the elders that grow down by our pond don't have a sweet floral perfume, but I find their scent beguiling nevertheless. It comes through in this unusual recipe, and the texture is unlike anything else I've tasted. Although *fritters* suggests a mass of fried dough aggregated around some sort of morsel, this version is lighter. The batter-dipped blossoms are fried more in the manner of a pancake than a deep-fried fritter. Start with just a tablespoon of oil in a nonstick pan, adding more as needed.

Makes six to ten fritters.

6 to 10 elderblow clusters
½ cup milk
¼ teaspoon vanilla extract
whites of 2 eggs
¼ teaspoon salt
2 tablespoons sugar
½ cup unbleached white flour
cooking oil as needed
maple syrup

Mix the milk, vanilla, egg whites, salt, and sugar. Stir in the flour. Pour a few tablespoons of oil in a nonstick skillet. Dip the flower clusters into the batter, and fry over low heat. Add more oil as necessary. Serve warm with maple syrup.

TEA AND ELDERBERRY SORBET

Try this sorbet as a midday, midsummer pick-me-up.

Serves four.
2 cups elderberry juice
2 cups brewed green tea
¾ cup honey
1 lemon, juiced

To make the elderberry juice, simmer 4 cups of elderberries in ½ inch of water for 5 minutes, crushing them with a potato masher as they soften. Pour into a strainer and press the pulp with the back of a large spoon.

Brew the tea. While it still is warm, stir in the berry juice, honey, and lemon juice until the honey is dissolved. Process this mixture in an ice-cream freezer, following the manufacturer's directions. Garnish the individual servings with mint leaves if you wish.

BOY SCOUT CHAMPAGNE

Also known as Southern Champagne, Elderblow Fizz, and Elderflower Cordial, this fresh, lightly fermented summer drink has all but been forgotten in North America. Some specialty shops import the sparkling nonalcoholic cordial in bottles.

Makes 3 quarts.
3 quarts water
2 cups sugar
juice and zest of 2 lemons
6 to 8 elderblow clusters

Harvest the clusters at their most fragrant. Over low heat in a soup pot, stir the sugar into the water until dissolved. Allow the liquid to cool, then add the lemon juice, zest, and blossoms. Press the blossoms down to submerge them. Put a cover on the pot and allow it to sit in the sun for a day or two. Expect the liquid to begin to ferment, in the manner of cider without preservatives. Strain through a colander lined with cheesecloth, then fill plastic soda bottles through a funnel. Twist on the tops just enough to allow the escaping carbon dioxide to hiss as the cordial continues to bubble. Refrigerate and serve chilled. The cordial can be mixed with seltzer water for a mild-flavored spritzer.

ELDERBERRY WINE

"Masculine" isn't a term you'd expect to find in the evocative text on a wine label, but that is how M. A. Jagendorf describes elderberry wine in his affable guide *Folk Wines, Cordials, and Brandies* (1963). The wine does have a forthright personality. As Andre L. Simon put it in *English Wines and Cordials* (1946), elderberry wine "had the reputation of being just like port, but only to look at, of course." There is a tradition in parts of the United States of using elderberry wine as a female tonic. Mothers would administer a couple of tablespoons of the homemade wine to ease their daughters' cramping.

Among folk wine enthusiasts, the elderberry stands out because it yields both a white wine (from the blossoms) and a red (from the berries). Raisins often find their way into recipes to add body and some sweetness. The spices and lemon can be omitted if you want the straightforward taste of the berries to come through.

If fresh fruit isn't available, you can substitute dried elderberries. Use from 4 to 6 ounces of dried berries per gallon of water. To lend color and a tannic zing to *other* berry wines, add 2 ounces or so of the dried elderberries per gallon. The fresh or dried elderberry blossoms can be added to this or any wine for that elusive floral scent. Dried berries, canned berries, and dried blossoms are available from wine hobby stores.

This recipe takes you only part of the way to the finished product. There is an enormous variety of wine-making methods, from folksy to forbiddingly high-tech, and you are referred to other books or a wine-making hobby store for an approach you'll be comfortable with.

8 quarts elderberries
2 gallons water
6 pounds sugar
1 pound raisins
1 ounce grated ginger
6 cloves
juice of 3 lemons
wine yeast

Place the water in a soup pot and bring it to a boil. Add the berries, sugar, raisins, ginger, and cloves and simmer for 20 minutes. Toward the end of the period, crush the berries with a potato masher. Strain the pulp through a colander lined with cheesecloth. Press the pulp with the back of a large spoon to express more juice. Stir in the lemon juice and yeast. And then the marvelous alchemy of fermentation will be on its way.

ELDERBERRY VODKA

The Danes have a tradition of steeping all sorts of fruits and herbs in neutral spirits to prepare what they call *akvavit.* In Copenhagen, Vivi Labo's many home recipes include an elderberry vodka. Danish folklore has it that the scent of elder blossoms could keep trolls from entering the home, and that this beneficial force can be added to alcoholic spirits.

Whatever its magical powers, elderberry vodka takes on complexity as it ages. Vivi notes the flavors of almonds, vanilla, and chocolate. The berries may improve with freezing, either harvested after the first frost or placed in the freezer. Defrost frozen berries in a bit of the vodka you'll be using.

Half fill a glass jar with the berries. Fill to the top with vodka. Allow the berries to macerate in a cool, dark place for several weeks, and give the jar an occasional shake. Pour though a colander lined with cheesecloth or a coffee filter. Bottle the flavored vodka for further aging.

CHAPTER 8

Gooseberries

Some habits don't translate from one culture to another, and the British fondness for gooseberries has never taken root in North America. Chances are most Americans go a lifetime without looking at a gooseberry bush, much less eating the fruit.

And yet, overlooking those nasty thorns for a moment, they are an excellent plant for the yard. The shrubs remain shrublike without much pruning, they have attractive leaves shaped like little gloved hands, and the ripe berries come in a remarkable range of colors, from sour-apple greens through yellow and orange to the deepest of plummy purples. The flavors vary as well, with overtones of other berries, grapes, and even apricots. Paler varieties tend to be less tart and less fruity. The deep reds are said to be best for pies.

So what's behind our resistance to this agreeable berry? One limiting factor might be that gooseberries have an identity problem. Because the flavor and appearance vary so much from one variety to another, there really is no defining gooseberry taste or color. Everyone knows precisely what a strawberry looks and tastes like and "raspberry" brings a certain hue to mind, but the gooseberry has always remained elusive. A curious confirmation of that is a tip from an 1884 gardening magazine, suggesting that readers could make a mock-gooseberry jam from seaweed and vinegar, among other homely ingredients.

Another complexity is that some gooseberry varieties customarily are picked when green and tart for use in cooking and in making jams. These are termed "culinary" gooseberries or "cookers," and include Invicta and Whitesmith. Other varieties, called dessert berries, aren't harvested until fully ripe, at peak flavor, and typically are yellow or red in color. Early Sulphur, Hinomaki Red, and Whinham's Industry are well-known examples. Returning to the thorns, they are a nuisance, but the British long ago put a positive spin on that. A quaint bit of folklore has it that vulnerable little fairies seek shelter from their enemies within those bristling branches.

GOOSEBERRY

And it was British gardeners who formed gooseberry clubs going back to the 1700s. By the following century more than 350 varieties were available, and these clubs competed annually to see who could grow the most colossal gooseberry. As gardening fascinations and fads go, gooseberries have had a terrifically long run; as late as the 1980s a few of these clubs still were around. A recent record-holder, by the way, was a Woodpecker gooseberry weighing in at over two ounces.

If you don't know much about gooseberries, you are in for a pleasant surprise—assuming you try a flavorful variety that has been picked at its prime. Some of the best-known gooseberries, Pixwell among them, may be the least worthy. "Pixwell is a little better than no gooseberry at all," said berry expert Ed Mashburn as we toured his plantings outside of Northumberland, Pennsylvania. Ed is a member of North American Fruit Explorers, a group that brings together amateur enthusiasts, researchers, and professional growers. He had kinder things to say about Hoenings Earliest, a large golden yellow berry with a taste that makes you wonder why some books write off gooseberries as material for pies.

GOOSEBERRIES IN THE YARD

Gooseberries don't need full sun and they will thrive in a cooler and somewhat shaded part of the yard. Because of their thorns, it's best not to place plants where younger children are apt to go scrambling for lost toys. When setting out the plants, prune the branches to just half their length to help make the shrubs stronger and more productive. There is a tradition of training gooseberries as a *standard*—that is, pruned so that a ball-shaped cluster of foliage tops a straight, branchless trunk. The result looks like a miniature tree. As the young gooseberry shrub develops, select one vertical stem as the trunk and rub off buds that form along the lower twelve to eighteen inches.

Gooseberries are relatively woody shrubs compared with the currants, bearing fruit on branches that are from two to four years old. So, in an annual spring pruning, you would be taking out only the old, darkened branches that have stopped producing. Allow a half dozen shoots to come up, snipping off the rest at the soil line.

GOOSEBERRIES IN THE KITCHEN

Gooseberries have to be picked with care, in a couple of senses. First, to avoid the prickles, hold the branches with a leather-palmed glove on one hand and pick the berries (or snip off berry clusters) with the other. Second, it's not all that easy to tell when gooseberries are ready to be picked. With most berries you can go by the color, but that is complicated by the gooseberry's range of colors and patterns when ripe. It's best to taste test from one day to the next to catch them at their optimal ripeness. If you have a number of different varieties, you might even take mug shots of berries as they look when at their peak flavor to keep for a reference in the following years. Keep in mind that the varieties considered to be "cookers" can be picked when not quite ripe for use in jams and recipes.

Before eating or preparing the fruit, pinch off the stem and the withered blossom end—"top and tail" them, as the phrase goes. You don't need to bother with this step if you will be cooking down the berries and straining them.

For interesting ways to make use of gooseberries, either get your hands on British cookbooks or surf the Internet for recipes offered by UK sites. The rich British pudding known as a fool often is made with gooseberries (see the recipe on page 29.)

WILTSHIRE WHITSUN GOOSEBERRY CAKE

Among the British foods traditionally prepared around the holy days of Whitsun, or Whit Sunday, are cakes that take advantage of the first fruits of the season. In the strawberry version, the cake may be topped with seven whole berries to symbolize the gifts of the Holy Ghost. This recipe uses gooseberries and the subtly fragrant white blossoms of the elderberry, which come into flower around the time the gooseberries begin to appear.

Serves six to eight.

2 cups tart green gooseberries
5 elderberry blossoms, chopped finely, plus whole blossoms to garnish
2 cups unbleached white flour
2 teaspoons baking powder
2 eggs

16 tablespoons butter, plus 2 tablespoons to coat pan
¾ cup sugar
confectioners' sugar to garnish

Preheat the oven to 350 degrees F. Snip off the elderflower clusters and shake off any insects. (If there aren't any handy elderberry bushes, you can substitute a handful of scented geranium or lemon balm leaves.) Finely chop the blossoms, leaving a few springs (or the substituted leaves) for a garnish. The gooseberries should be either a cooker variety or dessert variety picked less than ripe. In a mixing bowl, whisk together the flour and baking powder. Cut in the butter with two forks. Stir in the sugar and the eggs. Gently stir in the elderberry blossoms and gooseberries.

Use the 2 tablespoons of butter to coat the base of an 8-inch spring-form cake pan and then, after lining the pan with parchment paper, coat the paper as well. Pour in the cake mixture. Bake for 80 minutes or until an inserted toothpick comes out clean. Release the cake from the pan when cool. Sprinkle with confectioners' sugar and decorate with elderberry blossoms (or sweet geranium or lemon balm leaves). Serve the cake while it is still warm.

GOOSEBERRY WINE

Gooseberry wine has been popular in Great Britain for centuries—but then the British have been driven by their climate to make wine out of nearly everything that grows. So it is that gooseberries have suffered the second-string status of a substitute fruit.

The wine has a satisfactory body and good acidity, although sweetener has to be added to develop a percentage of alcohol within the usual range. Alternately, winemakers have added neutral distilled spirits to their gooseberry vintages. According to a British recipe from *A Philosophical and Statistical History of the Invention and Customs of Ancient and Modern Nations in the Manufacture and Use of Inebriating Liquors,* published in 1838, gooseberries are picked when half ripe, then crushed and strained. To each gallon of the resulting juice, add 3 cups of sugar. The wine is considered ready to drink in about six months. You still can find gooseberry wine in the Isles; in fact, Great Britain's northernmost winery, Orkney Wine Company, manages to coax wines from Scotland's gooseberries, as well as from elderberries and black currants.

WILD RED RASPBERRY

CHAPTER 9

Raspberries

Beyond their enchanting flavor and aroma, raspberries seem to have a certain personality as well. Thoreau described them as innocent. Others find them to be cheering, even mood elevating, so that they can act as a comfort food even for people who have no childhood or cultural associations with them. The raspberry's essence is amazingly complex. An Oregon State University study of its scent had to take seventy-five aromatic compounds into consideration.

Fresh-picked raspberries are a fragile delicacy, and they take on a remarkable richness when cooked down in a jam. The French make a red raspberry liqueur, crème de framboise, and one of black raspberries, marketed under the brand name Chambord.

Raspberries haven't received as much attention as certain other berries by researchers looking for healthful properties in fresh foods. Raspberry-leaf tea has some reputation as a women's tonic; try picking the new leaves just as they are opening and add mint leaves to improve the flavor. It seems the leaves do something for horses, too; a New Hampshire company sells the air-dried leaves in bulk for veterinarians, who use the herb to sooth pregnant mares.

BLACK RASPBERRIES

Their color suggests that black raspberries have a depth of flavor beyond that of the red berry, and they do have an aroma and a slight muskiness all their own, possibly bringing black cherries to mind. In recipes, they sometimes are preferred over red varieties because their flavor tends to hold up better when cooked and in the company of other ingredients.

In drinks and desserts, black raspberries contribute a lovely lavender shade that is at once warm and cool. The berries have been used in processed foods as a natural colorant, rather than resort to a chemical product with no nutritional benefit. They once supplied ink for the familiar purple USDA stamp on meat and to tint Dr Pepper soda.

Purple raspberries are a cross between red and black varieties, with a flavor closer to the red parent.

YELLOW RASPBERRIES

The yellow raspberry is a striking variation on the red form, with a milder taste. They are even more crumbly and evanescent once picked. Yellow raspberries can be used to stunning effect when arranged in a pattern with red or black berries atop a cheesecake.

WINEBERRIES

Raspberries glisten like jewels, but wineberries do them one better. They glow like taillights in a traffic jam, as if illuminated by tiny red-orange LEDs. The fruit is protected within a furry husk that opens as the berry ripens, and the canes bristle with stout red hairs. While the sprightly flavor isn't quite as sweet or aromatic as that of fully ripened raspberries, wineberries have their own pleasant character. The species was brought over from Asia and yet seems to fit in comfortably with the landscape, as they have been around as long as their neighboring blackberries and raspberries. We find that our hedgerow plants bear generously and dependably, and we anticipate them with pleasure each year. Although the canes do look formidable, the hairs aren't all that scratchy, and we allow a few wineberries to spring up in flower beds and alongside outbuildings.

SALMONBERRIES

The beautifully colored salmonberry grows wild along the Pacific Coast and into the Rockies. The berries are fragile and mild in flavor, with a modest yield (although a 1985 study reported in the journal *Ecology of Food and Nutrition* claims it is possible to pick 0.27 quart of salmonberries in five minutes). Together, these qualities have kept the fruit from being produced commercially. In fact, you'll have to do some searching to find a source for a few plants to set out in the yard, even though the large pink flowers are standouts among berry blossoms. Coyotes, black and grizzly bears, possums, grouse, and quail are among the animals attracted to the berries, and beavers browse the foliage.

There is no one certain explanation for the species' name. The blossoms appear when the salmon spawn; the berries ripen as salmon are running western rivers; the color of the ripe fruit ranges from a warm yellow to reddish, with a salmon tone falling somewhere in the middle; and most intriguing, an infusion of the bark is said to bring relief from eating too much salmon.

THIMBLEBERRIES

The thimbleberry is a native raspberry relative, found growing from northern Michigan to the northwestern United States and British Columbia. The large white flowers are especially attractive and the fruit has a good flavor. The shoots are edible, too, if you're in a pinch; Native Americans in what is now British Columbia would peel and eat them. The plants don't produce a lot of fruit, however, and thimbleberries rarely are considered for the home landscape. In Michigan's Upper Peninsula, at least one small company picks enough of the scarce berries to market a thimbleberry jam ("Ingredients: Thimbleberries, sugar, and many hours in the bush," reads the label from Thimbleberryjam.com).

RASPBERRIES IN THE YARD

Raspberry plants come armed with thorns, of course. They are aggressive growers, too—a raspberry plant left untended will be inclined to create a formidable thicket. If you want something guaranteed to not scratch or overtake a small yard, you're better off with blueberries or currants.

Although raspberries can reproduce themselves by seed, they spread eagerly by sending up new canes at an impressive (and sometimes inconvenient) rate. Red and yellow varieties produce canes from their roots. The canes of black raspberries droop to the ground, and as the tips make contact with the soil they are stimulated to sprout new growth. Purple raspberries are a cross of the two types, and they share both growth habits.

As new canes are produced, the old ones die, producing fruit for just one growing season. That means you've got to face those thorns and prune in order to keep plantings vital.

Raspberries typically are grown in rows with the support of a trellis—wires suspended between posts like dwarf telephone poles. If the wires are to be taut enough to support the canes without a discouraged sag, the uprights must be anchored securely. That takes work. You've got to dig holes a good two feet deep, then fill them with either a couple of bags of concrete mix or, with less trouble, a combination of gravel and packed soil. The posts still may want to lean toward each other, and you can keep them upright by extending the wire from either end of the trellis to an earth anchor—a large metal object that is screwed into the earth like a wine opener into a cork. A hardware store or home center will stock most of the parts you'll need to build and install a trellis, but you may have to shop via mail order to buy specialty items such as 10.5-gauge trellis wire, wire vises (to grip the ends of the wires), turnbuckles (to tighten wires once they are installed), and earth anchors. In New York State, gardener Pooniel Bumstead chooses not to bother with wires. She runs hemp twine between a row of cedar posts, and these supports can be removed along with the old canes when the harvest is done.

In deciding which raspberries to grow, you have a choice between varieties that produce one crop a year and the so-called *everbearers* that offer the bonus of a second flush of fruit in the fall. Then consider how cold your winters are apt to be. Red and yellow raspberries rate as quite hardy; black and purple varieties are somewhat less so. Everbearing raspberries may produce their fall crop too late to avoid frost damage in northern states. If you'd rather not have just one mad flurry of berry picking each summer, purchase varieties that ripen at different times in the season. Or just select plants of the various colors—red, yellow, black, and purple—and you'll have crops coming in over a longer period.

Set out bare-rooted plants in either spring or fall; if you're using potted plants, they can go in during summer months as well. You'll pay more for plants in pots, but they'll be sturdier from the outset and put out more new growth. Bare-root stock should be pruned back to just a couple inches above ground level; there's no need to prune plants from pots unless you see damaged canes.

Plant the raspberries so that you'll be able to get to them for regular maintenance—pruning and mulching, and mowing if they are to be surrounded by grass. Set red and yellow varieties two feet apart and purple and black raspberries four feet part. Whichever type of berry you are growing, lay out parallel rows roughly eight feet apart.

CULTIVATED BLACK RASPBERRY

You can extend your berry beds without buying nursery stock by taking advantage of the new plants that spring up from the roots or cane tips. Give a jab with a spade to sever the little plant, taking care to exhume its roots, and then transplant it to a new location.

Although raspberries will manage in indifferent soil, you'll get better results if you add manure and compost when turning the beds. In the years to come, fertilize in early spring with manure or a commercial product in the ratio of 10-10-10. Weeds and grasses aren't afraid of thorns, and they soon will carpet the berry patch unless you intervene. Control them with a generous layer of mulch—leaves, shredded bark, or straw—placed over a barrier of newspaper. Or, if you don't have a supply of these materials on hand, you can lay down rolls of porous landscaping cloth or black plastic. Alternately, mow closely along the berry rows.

PRUNE THEM OR BE SORRY

Raspberries, if left to their own devices, make a mess. The canes they send up so lavishly will put out fruit a single season; after that they become deadwood useful only to nest-building birds. Eventually, the berry planting can die out. So, each year after the harvest, you'll need to pull on a pair of stout gloves, get out the pruning shears, and go after the pale, silvery, brittle ghosts of the old canes. Then turn your attention to the green, living canes that will produce the next crop. Prune away those that look particularly spindly or injured in any way—raspberries benefit from a firm hand when pruning. In established rows of red and yellow raspberries, you also should remove living canes until there are no more than three to five for each running foot of row, as well as those that are straying from the rows.

For everbearing raspberries, note that the fall crop is produced on canes that came up that spring, rather than on the previous year's canes as is typical of other raspberries. Once you've harvested the fall's berries, you can prune off the top section that yielded the fruit, allowing the remaining part of the cane to produce the next summer's crop.

Black raspberries need a variation on this regimen because of their habit of growing until they flop over and take root at their tip. This is nature's way of making more plants, but if you pinch off the end four inches of new canes, the plants will put their energy into sending out fruit-bearing lateral branches. After the harvest, take out the older canes that have fruited. The following spring, remove canes to leave just five or six per plant, trim the lateral branches to one foot or so, and tie the canes to the trellis wires.

Wineberries propagate themselves like black raspberries. Their canes grow until they curve back down to earth, where they sprout in the soil. This is fine in a wild berry patch, but to rein in a home planting, prune each fall.

THE HARVEST

Pick them as they become ripe, handling the berries gently and using small containers to prevent crushing. Underripe fruit won't get any better once it is picked. Store the berries in the refrigerator for no longer than a couple of days, ideally at a temperature just above freezing. Or keep their goodness indefinitely by freezing them or converting them to jam or syrup.

RASPBERRIES IN THE KITCHEN

Back in the 1800s, families with even a modest yard would grow raspberries for eating fresh and putting up as jam. Today, we can enjoy the convenience of buying raspberries flown in from wherever on the globe summer is taking place. But, consider: are we giving up something in the bargain? Do those intercontinental berries still have the innocence that Thoreau referred to? Or, less abstractly, do they taste very good? One reason that fresh out-of-season berries are so expensive is that a considerable percentage of them become worthless in transit. To increase the chances that raspberries will survive a long trip, growers harvest them when not quite ripe. So, what you are apt to buy in those pricey little plastic containers are firm, good-looking berries that didn't have the chance to develop their full flavor on the cane.

Try a taste test for yourself. Sample a few berries right out of the container to determine if they have an insipid flavor or a chemical aftertaste. If so, you might want to either buy frozen berries or wait until the local crop comes in. And as a last-ditch option, you might try a recipe for mock raspberry jam that's been circulating the Internet. You chop up green tomatoes to the size of raspberries, open a box of raspberry-flavored gelatine dessert—or, no, better restrict yourself to the first few options.

If there is one less-than-perfect item in the raspberry's résumé, it is the seeds. They may strike you as an agreeable bit of substance in a jam or tart, but if not, remove them at the outset of a recipe with a colander lined with cheesecloth or by using a steam juicer. There's a middle path, too: a sieve will allow some pulp and a perhaps a few seeds to pass through, depending on the gauge of the mesh.

Wineberries can stand in for raspberries in recipes, but you may want to bump up the sugar a notch.

RASPBERRY RHUBARB RELISH

Here is a summertime variation on the standard Thanksgiving relish of cranberries and entire oranges.

Makes about 1½ quarts.

6 cups rhubarb, chopped finely
3 cups sugar
2 oranges
2 pints raspberries

In a mixing bowl, combine the rhubarb and the sugar and allow the mixture to stand overnight. Simmer the mixture without adding water, stirring to help dissolve the sugar unless the oranges are organic or from your own tree. Wash them in warm soapy water to remove any chemicals or preservative coatings from the skin. Cut them in eighths and remove the seeds. Place all of the ingredients in a food processor and pulse briefly to make a chunky puree. Allow the sauce to spend a night in the refrigerator before serving.

BAKED RICOTTA WITH RASPBERRY COULIS

This recipe tastes like cheesecake, with less fuss and less cholesterol. Ricotta cheese bakes down into a firmer, more interesting version of itself. Because you are using just a small area of the oven for a considerable time, you might take advantage of the heat to also bake bread, rolls, or focaccia to accompany the ricotta. The coulis can be made from fresh or frozen raspberries of any color.

YELLOW RASPBERRY

WINEBERRY

Serves eight.
1 pint ricotta
1 pint fresh or frozen raspberries
¼ cup water
2 tablespoons Cointreau or Grand Marnier
zest of 1 lemon

Preheat the oven to 375 degrees F. Oil the inside of a 9-inch pie pan or quiche dish. Shape the ricotta into a low, flat mound so that its edges don't quite extend to the sides of the pan or dish. Bake for from 40 minutes to 1 hour and 20 minutes—long enough that the top begins to turn golden brown and the ricotta is firm enough to come free when lifted with a spatula. Low-fat ricotta will require more time in the oven.

Place the raspberries and the water in a saucepan and simmer for 10 minutes, stirring occasionally. Crush the softening berries with a potato masher. For a seedless sauce, place the berries in a strainer lined with cheesecloth; gather the edges of the cloth to make a ball and squeeze to express as much juice as possible; and return the juice to the saucepan. Add the liqueur and lemon zest, and simmer for another 5 minutes.

The coulis can be poured over the top of the baked ricotta mound in a decorative spiral or lattice. Or just spoon the coulis over each wedge-shaped serving of ricotta.

RASPBERRY CHOCOLATE MOUSSE

At a recent New Year's party, it struck me as unusual that neither the hosts nor the guests showed interest in the champagne that was languishing in the refrigerator. Instead, attention hovered around a buffet of attractive dishes, in the center of which was an enormous glass bowl of this mousse. As I discovered when I later called my friend Gail Stern for the recipe, it's a big-bore dessert, with real whipped cream and real chocolate and plenty of eggs (raw ones, at that). All of those ingredients might come across as heavy—well, they *are* heavy—but the addition of fresh raspberries and crème de framboise somehow has a leavening effect. The result? You can either eat more *or* feel a bit better about what you already have eaten.

Serves ten to twelve.

12 ounces bittersweet chocolate chips
1 pint whipping cream
3 tablespoons sugar
6 eggs: 6 whites and 2 yolks
6 tablespoons crème de framboise or Chambord
1 pint fresh raspberries

Melt the chocolate in a double boiler. Whip the cream with the sugar to form stiff peaks. Separate the eggs. Whip the egg whites. In another bowl, whisk the two yolks with the melted chocolate and the liqueur. Fold the yolk mixture into the whites. With a few strokes, fold this egg mixture into the whipped cream along with two-thirds of the berries; allow light and dark streaks to show rather than overmixing. Put the mousse in a serving bowl and decorate with the remaining berries. Refrigerate the dessert until you are ready to serve it.

LINZER TARTLETS

Smaller than the standard Linzer tart but larger than cookies, these tartlets are the right size to serve individually for dessert. They have the traditional latticework on a reduced scale. Linzer recipes usually call for a filling of raspberry jam, but the stuff can turn tough and chewy in the oven and this version uses a homemade raspberry sauce. If you don't have a 4-inch-diameter cookie cutter, you can make a simple cardboard pattern and use a sharp knife (I favor an X-Acto craft knife) to cut the rounds for the tartlet bases. Use an edge of the cardboard to guide you in cutting straight lattice strips.

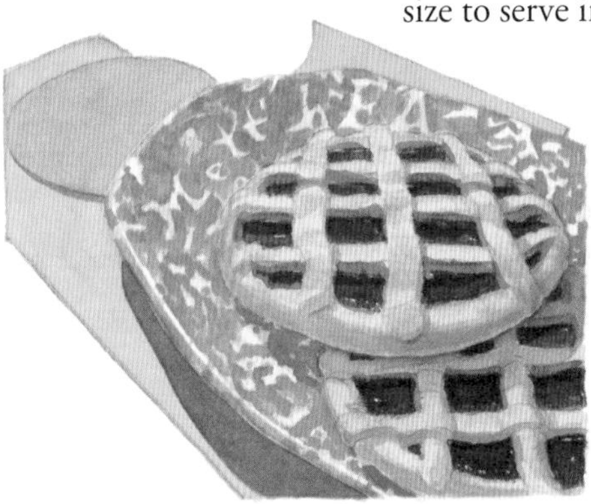

SALMONBERRY

Makes seven or eight tartlets.

For the jam:

1½ cups fresh or frozen raspberries
¼ cup water, plus 1 tablespoon to dissolve the cornstarch
½ cup sugar
1 tablespoon cornstarch

For the dough:

¾ cup raw slivered almonds
⅔ cup sugar
1 stick of butter (4 ounces)
2 eggs
zest of 1 lemon
1 teaspoon vanilla extract
1¾ cups unbleached white flour
½ teaspoon ground mace
¼ teaspoon ground cloves
¼ teaspoon salt
confectioners' sugar

With the ¼ cup of water in a saucepan, simmer the berries for 5 to 10 minutes to help release their juice. Crush the berries with a potato masher as they soften, then strain them through cheesecloth. Gather the edges of the cheesecloth to form a ball, then squeeze to express as much juice as possible. Pour the juice into the saucepan, add the sugar, and simmer with an occasional stir until the sugar is dissolved. Dissolve the cornstarch in 1 tablespoon of water, and stir this into the juice. Place the pan in the refrigerator to speed the thickening.

Preheat the oven to 350 degrees F. Toast the slivered almonds in a countertop oven until golden brown. Allow the almonds to cool, then place them and the ⅔ cup sugar in a food processor with a metal blade and pulverize the almonds into a fine meal.

In the food processor with a paddle attachment, beat the butter until softened. In another bowl, whisk the eggs along with the lemon zest and vanilla extract. Add this mixture to the butter and beat again. Add the flour, mace, cloves, and salt, and beat just enough to form dough. Divide the dough into two balls, one of them roughly twice the weight of the other. On a lightly floured surface, flatten the balls by hand and then with a roller, forming discs about ½ inch thick. Wrap the discs in plastic wrap and place them in the freezer to cool for 15 minutes or so.

Unless you have a cookie cutter that will make 4-inch-diameter rounds, cut out a circular pattern from a sheet of single-ply cardboard (such as the backing of a pad of paper). Draw a circle by tracing the bottom of an appropriate-sized can or use a compass. Cut out the circle and save the rest of the cardboard sheet for use as a straightedge, as shown in the illustration.

Remove the chilled dough from the refrigerator. Using just enough flour to prevent sticking, roll the discs to an even thickness of ¼ inch. Make seven or eight bottom circles from the larger disc by cutting around the cardboard circle. As necessary, gather the scraps, form them into a ball, roll, and cut more discs. For the lattice, cut ⅜-inch-wide strips from the smaller disc. The rim around the perimeter of each tartlet is the only tricky step—it should bond well with the bottom to prevent the raspberry filling from leaking out. Think in terms of a backyard wading pool. To help the rim adhere, dip a finger into a glass of water, run it around the disc's edge, then place strips around the edge *floured side up.* Press down lightly to make a good bond.

Place the discs on a cookie sheet. Spoon raspberry jam into each.

For the latticework, you'll need to cut six strips for each tartlet. There are two long center strips placed at right angles to each other, and both are flanked by a shorter strip on either side. Don't be too concerned with securely attaching the latticework. Again, you may need to gather the scraps into a ball and roll the leftover dough to make more strips.

Bake for 15 minutes or until the tartlets just begin to turn golden around the edges. Sprinkle confectioners' sugar over them while still hot so that it will be visible only on the latticework, not on the jam; use a sieve to ensure an even dusting. Before any leaked filling can bond the tartlets to the cookie sheet, place them on a wire rack to cool.

RASPBERRY GRATIN

You might not expect to find them on a dessert menu in Bangkok, but raspberry gratin with crème de framboise ice cream (see the following recipe) recently was offered by the Thai city's Le Normandie Restaurant. Try this warm-and-cold duo some sultry summer night, wherever you happen to be. The gratin recipe is adapted from one used by the late Joan Sturm of the Marin County (California) Farmers' Market.

Serves four to six.
1 tablespoon butter
5 tablespoons sugar
1 egg
⅓ cup milk
⅓ cup unbleached white flour
½ teaspoon salt
zest of 1 orange
3 cups raspberries
2 tablespoons coarsely chopped almonds

Use the butter to coat an 8- or 9-inch gratin dish or other shallow baking dish. Sprinkle 2 tablespoons of the sugar over the bottom of the dish. Whisk together the egg, milk, flour, salt, and orange zest and pour this batter into the dish. Spread the raspberries evenly over the batter, pressing down with a spatula to set them deeper. Sprinkle the remaining 3 tablespoons of sugar over the top. Bake 12 to 20 minutes, or until the batter is set, and serve warm with the ice cream.

CRÈME DE FRAMBOISE ICE CREAM

Any berry liqueur or syrup can be used in this recipe. You might want to include ¼ cup of whole or mashed berries as well.

Serves eight to ten.
1½ cups whole milk
½ cup sugar

1½ cups heavy cream
¼ cup crème de framboise or Chambord
¼ teaspoon vanilla extract

In a mixing bowl, whisk together the milk and sugar until the sugar is dissolved. Stir in the cream, liqueur, and vanilla extract and process the mixture in an ice-cream freezer according to the instructions.

PINK RASPBERRY LEMONADE

Mix raspberries with lemons for a lemonade that is naturally and flavorfully pink.

Serves four.
2 cups berries
4 cup water
1⅓ cups lemon juice
4 tablespoons sugar

Puree the raspberries with the water in a blender or food processor. Remove the seeds with a sieve if you wish. Add the lemon juice and sugar and stir well until the sugar is dissolved.

RASPBERRY SHRUB

When life gives you lemons, make lemonade. So goes the well-worn saying. A kitchen corollary might be: if you have lemons, you can go easy on the vinegar. As pleasing as vinegar is in a salad dressing, it has a funky edge in an old-fashioned *shrub,* the vinegar-and-fruit beverage popular before tart citrus fruits were readily available across North America. The term has nothing to do with small bushes, but is derived from the Arabic *sharab,* meaning wine or beverage. This recipe uses lemons, with only a jot of vinegar as a nod to history. Aside from being a drink base, berry shrubs can be used as an ingredient in marinades and glazes, salad dressings, simmered sweet-and-sour red cabbage, and mixed drinks of your own imagining.

Makes about 1 quart.
4 cups raspberries
2 cups water
juice of 2 lemons
2 teaspoons raspberry, wine, or cider vinegar
¾ cup sugar

In a saucepan, simmer the berries in the water for 5 minutes or so. Crush the softened berries with a potato masher, then strain the berry mixture through cheesecloth placed in a colander. Press with the back of a large spoon to yield as much juice as possible. In the saucepan, simmer the raspberry juice with the lemon juice, vinegar, and sugar for another 5 minutes, stirring until the sugar is dissolved. To make a shrub, use 3 tablespoons of this mixture in a tall glass with ice and either cold seltzer or still water.

SIX-FOUR-TWO COCKTAIL

The proportions of this French cocktail are stated in the name.

Makes four cocktails.
6 ounces raspberry juice, sweetened to taste
4 ounces dry champagne or sparkling white wine, chilled
2 ounces crème de framboise

Half fill a highball glass with ice cubes. Pour in the ingredients, taking care not to agitate the champagne so that it loses its fizz. Give a slow stir and serve.

RED TULIP

This is a lovely liquid confection for those who don't care for dry cocktails.

Makes one cocktail.
1 ounce Madeira
1 ounce crème de framboise
1 ounce Cointreau or Grand Marnier

2 dashes Angostura bitters
lemon twist for garnish

Chill a champagne flute. Stir the ingredients in a shaker half full of ice and serve straight up with the lemon twist.

CHAMPAGNE FRAMBOISE (KIR IMPERIALE)

Champagne cocktails have been around forever. They are best enjoyed if you have a light hand with the amendments.

Makes one cocktail.
1 ounce crème de framboise
5 ounces champagne or sparkling dry white wine, chilled

Chill a champagne flute. Pour in the crème de framboise and two fingers of champagne, give a stir, and fill with champagne.

DANISH FRUIT PUDDING WITH CREAM (RØDGRØT MED FLØDE)

See the following chapter for this recipe, with its blend of raspberries and red currants.

RED CURRANT

CHAPTER 10

Red Currants

Red currants are brilliantly transparent and hang in glorious pendants from shrubs of a backyard-friendly size. European varieties were brought over to colonial Massachusetts in the early 1600s, and the berries became a common sight around American homesteads.

And yet red currants have slipped from favor in the United States. They lack the sweetness and fragrance of raspberries and strawberries, and their seeds get in the way of fully enjoying their sprightly tartness. The berries are primarily used to make jelly, puddings, and juice, rather than being eaten fresh. Recently imported varieties may change that, however. Tartran, illustrated here, is a marked improvement. The berries are larger than usual, with relatively negligible seeds and a pleasantly sweet flavor. You'll have to be patient if you want to sample Tartran, because it may be some time before the variety is grown in the United States on any scale.

White currants are as lovely as any berry or fruit, looking like miniature crystal balls with which it might be possible to divine a few moments of the future. They are less tart than the red berries. The *pink currant* is an eye-catching variation. (See Chapter 3, Black Currants, to read about this currant of a different color and dramatically distinct flavor.)

RED CURRANTS IN THE YARD

Red currant shrubs have a compact, easily maintained form that suits the yard and garden. Unlike their relative the gooseberry, they don't brandish thorns and the berries grow in clusters rather than individually. The plants do well in regions with cool summers and may benefit from dappled shade or a northern exposure in warmer areas. Set them three to five feet apart, or somewhat closer for a hedge. Because these currants don't have to cross-pollinate, they can be grown alone in various spots around the yard. Prune them early each spring to leave no more than a half dozen of the new branches that emerge from the ground.

RED CURRANTS IN THE KITCHEN

Many currant recipes are a legacy from northern Europe, where the berries continue to be appreciated for their lively taste and beauty. Typically, the seeds are removed from the berries, although there is no harm in eating them. If you are adding just a few currants to a green or fruit salad, the seeds may not be all that noticeable.

You can use your fingers to rake currants off the strigs while in the garden, but you may find it more convenient to snip off the clusters and remove the individual berries indoors. Refrigerate just-harvested berries for up to two weeks, either on the strig or off.

Red currants contain a considerable amount of pectin and are a traditional favorite for making jelly.

GRANDMOTHER'S APPLE CAKE (BEDSTEMØDER'S ÆBLEKAGE)

It's not really a cake as such but my Danish-American grandmother, Theodora Mathiasen Yepsen, really did make it, and often. Her traditional recipe included an untraditional ingredient, zwieback, the dry toast that has been sold as a child's teething aid for over a century. You also can use bread crumbs, either home-prepared or ready made. The dabs of slightly bitter red currant jelly serve as a counterweight to this dessert's sweetness. (Be aware that inexpensive red currant jelly may have little flavor.) Cranberry or lingonberry jelly will do just as well.

Serves six.

1 cup zwieback crumbs
4 tablespoons butter
½ teaspoon salt
2 tablespoons sugar
½ teaspoon ground cardamom
½ teaspoon ground mace
3 cups applesauce
1 cup (8 ounces) red currant jelly
whipped cream as topping

WHITE CURRANT

Preheat the oven to 375 degrees F. Convert the zwieback into crumbs by going over the toasts with a rolling pin. Melt the butter in a skillet over low heat, and sprinkle on the salt, sugar, cardamom, and mace. Add the crumbs and stir until they are golden brown.

In a one-quart baking dish, make the first of three or four layers of crumbs. Add a layer of about ¼ cup of applesauce, and dot with five or six teaspoons of the jelly. Continue making layers until you've used all the ingredients, ending up with a top layer of crumbs. Dot a final time with jelly. Bake for 20 minutes. Serve warm, topping each portion with whipped cream; or refrigerate and serve cool at a later time.

DANISH RED PUDDING WITH CREAM (RØDGROT MED FLØDE)

My grandmother would make this simple fruit pudding when she came to visit. It typically is made from a blend of berries, most often red currants and raspberries for a flavorful sweet-tart balance. If you substitute black currants for the red, the pudding will take on a greater depth of flavor and a beautiful dusky maroon hue. When fresh currants aren't available, you can substitute 1 cup of currant jelly for this fruit, reducing the sugar by half or to taste.

Serves four to six.

1 pint red currants
1 pint raspberries
½ cup sugar
2 tablespoons cornstarch
4 tablespoons water
1 cup whipping cream

Place the berries in a saucepan with barely enough water to cover them and simmer for 10 minutes. Crush the softening berries with a potato masher and strain through a colander lined with cheesecloth. Gather the edges of the cheesecloth to form a ball to squeeze as much juice as possible from the berries. Pour the juice into the saucepan, bring it to a simmering boil, and stir in the sugar until dissolved. Mix the cornstarch with the 4 tablespoons of water, then stir into the berry juice. Continue stirring for 3 minutes. Pour into glass dessert dishes and refrigerate until you are ready to serve the pudding. Whip the cream and top each portion with it.

WHIM WHAM

That's right, Whim Wham. The name is from a Scottish term for a whimsical fancy or a trifle—and *trifle* happens to be another term for this dessert. With its ladyfingers and cream, Whim Wham comes across as a simpler, lighter version of tiramisu. If you use packaged ladyfingers (or small slabs cut from store-bought angel food cake), the recipe can be put together quite quickly. The one time-consuming twist is to prepare the frosted black currants, if you choose to top the dessert with these little sweet-tart gems.

Serves eight.

1¼ cups heavy cream
2 tablespoons sugar
2 tablespoons Lillet Blanc aperitif
4 tablespoons crème de cassis or black currant syrup
zest of 1 lemon
¼ teaspoon salt
24 to 36 ladyfingers

For frosted currants, optional:

25 to 30 black currants
white of 1 egg
1 tablespoon water
¼ cup sugar

Allow time to make the frosted currants if you will be using them; see the directions that follow. Combine the cream, sugar, Lillet, crème de cassis or black currant syrup, lemon zest, and salt. Beat until the mixture forms peaks. Build the Whim

Wham in a serving dish at least 9 inches in diameter and deep enough to allow layering the ingredients. Begin with a layer of one-fourth of the cream, followed by one-third of the ladyfingers, and then one-third of the red currant jelly. Smooth the jelly as much as possible to make an even layer. Continue making layers, ending up with one of whipped cream. Keep the dessert in the refrigerator until you are ready to serve it.

To make the frosted currants, briefly whisk the egg white in the water, and pour in the berries. Stir to thoroughly wet the berries, then place them in a colander or sieve to drain for 5 or 10 minutes. Sprinkle the sugar on a dinner plate and scatter the berries over it, rolling them until covered with sugar granules. Pour the berries onto paper toweling and allow them to dry for at least 2 hours. Place them on top of the dessert just before serving.

CUMBERLAND SAUCE

This is a traditional British accompaniment to roasted meat but it also can be used as a marinade when grilling fish, tofu, or vegetables. Try the sauce with any sort of cheesy baked vegetarian dish. Another good pairing is Cumberland sauce ladled over blocks of tempeh that have been sautéed in olive oil.

The sauce has a fruity base of wine and of red currants. A variation uses elderberry wine instead of port; if you have some of the wine on hand, bring it up to the sweetness and alcoholic proof of port by adding sugar to taste and ¼ cup of brandy.

Yields about 2½ cups.

1 cup ruby or tawny port
1 cup red currant jelly
3 tablespoons molasses
zest and juice of 3 lemons
zest of 3 oranges and juice of 1 orange
1 tablespoon grated fresh ginger
1½ teaspoons prepared mustard
1½ teaspoons hot sauce

1 teaspoon salt
2 tablespoons cornstarch
4 tablespoons water

Place all of the ingredients in a saucepan *except* the cornstarch. Simmer with an occasional stir for 15 minutes. Thoroughly dissolve the cornstarch in the water and stir this mixture into the sauce. Continue simmering another couple of minutes. The sauce can be refrigerated for up to a few weeks. Consider freezing half the batch for later use.

BAR LE DUC PRESERVES

There have been white currants growing on our farm since we moved in more than twenty years ago, but the shrubs weren't producing last year and I wasn't able to try this most delicate of currant recipes.

Bar le Duc preserves originally came from the French town of that name, in Lorraine. They are produced from white currants, the tiny seeds of which are removed by a delicate operation. Make a tiny incision in one side of each currant with embroidery scissors and pluck out the seeds with a needle (or the quill of a feather, as was the tradition), being careful to retain the berry's shape. Weigh the seeded berries and measure out the equivalent of a light-flavored honey. Gradually bring the honey to a low boil in a saucepan, then gently add the berries. Simmer for 3 to 5 minutes. Remove the berries with a slotted spoon, and continue to simmer the honey until it thickens to the desired consistency. Place the berries in glass jars, pour the hot honey over them, and seal the jars.

ROSE HIPS

CHAPTER 11

Rose Hips

Rose hips are little nodes of nutrition, packed with vitamin C. Ounce for ounce, hips have fifty times the C of oranges, which might sound improbable if you've never popped a rose hip in your mouth. You *can* eat the hips right off the bush, but they are seedy and too tart for most tastes.

Still, they make an agreeable addition to herbal teas. And if allowed to go unharvested, they offer a bit of color in the winter landscape. To bring those dabs of scarlet indoors, snip the hips with a bit of branch and add them to cold-weather bouquets of dried flowers and grasses.

ROSES IN THE GARDEN

You can select rose varieties that have especially generous-sized hips (some varieties have none at all). These include the species roses *Rosa rugosa,* which need little care, and *R. moyessi* with its crimson blooms. The hips of Frau Dagmar Hastrup are said to be particularly high in vitamin C. Note that shrubs won't produce hips if you routinely deadhead them. Also, avoid spraying pesticides on shrubs from which you will be harvesting hips for culinary use.

ROSE HIPS IN THE KITCHEN

If you can put off picking rose hips until after the first or second frost, they will be at their sweetest. Halve them and allow them to dry indoors. When they begin to look shriveled, carefully remove the seeds and hairy pith. Allow the hips to dry further, then store in plastic bags in either the refrigerator or freezer.

ROSE HIP SYRUP

A tablespoon of rose hip syrup in hot water makes a pleasant, lightly flavored tea. During World War II, citrus fruits became so scarce in the United Kingdom that the British Ministry of Food issued directions on how to harvest hedgerow hips and make a syrup rich in vitamin C.

Makes about 1 pint.
4 cups rose hips
2 cups water
1 cup sugar

In a saucepan, simmer the hips in the water for 15 minutes. Pour the water and hips through a colander lined with cheesecloth, pressing the pulp with the back of a large spoon. To make sure that the little hairs don't end up in the finished product, you can pour this liquid through a coffee filter. Bring it to a simmering boil and stir in the sugar until dissolved. Allow the syrup to cool and bottle it.

ROSE HIP TEA

Rose hips lend tartness and color to an herbal tea. They are available dried from natural food stores. Or you can prepare your own by following the directions given earlier for cleaning and drying the hips. To reduce the brewing time, grind them in a coffee grinder or food processor. For starters, use 1 teaspoon per serving and sweeten to taste with sugar or honey.

ROSE HIP SOUP

If you aren't a fan of the thin, acidic taste of rose hips in tea, you might wonder why anyone would make soup out of the things. But simmering transforms the hips, bringing out a honeyed, plumlike aroma. This recipe calls for honey as a sweetener and a bit of the honey-based liqueur Drambuie as well.

You may be able to find loose-packed dried rose hips at a natural food store. If not, remove the crushed hips from bags of pure rose hip tea; you'll need about 15 bags to make the 5 tablespoons called for in the recipe. Or process hips as described earlier in the section "Rose Hips in the Kitchen."

Serves four.

1 quart water
5 tablespoons dried rose hips or fresh rose hips, minced
2 tablespoons honey
1 tablespoon lemon juice
1 tablespoon balsamic vinegar
½ teaspoon tamari (or salt to taste)
2 tablespoons Drambuie
yogurt or sour cream

Simmer the hips in the water for 10 minutes and pour through a strainer. Gather the edges of the cloth to squeeze the remaining pulp and express as much juice as possible. Add the honey, lemon juice, vinegar, tamari or salt, and Drambuie. Simmer for 5 minutes, stirring occasionally. Serve lukewarm or at room temperature. Stir in a swirl of yogurt or sour cream.

WILD STRAWBERRY

CHAPTER 12

Strawberries

The plump, cheerful strawberry is North America's favorite for all the right reasons. It is a bright berry red. The flavor is sumptuous. Its fragrance can travel a great distance through the air. The seeds are negligible. And for the past century and a half, strawberries have been available in a crowd-pleasing jumbo size.

Native Americans made do with the tiny wild berry, eating it fresh and adding it to cornmeal. For the Iroquois, the early summer appearance of the berry heralded the coming of bountiful crops, and strawberries were gathered for a thanksgiving feast. When European settlers arrived, they were impressed by the profusion, if not the size, of what they named the Virginia strawberry (*Fragaria virginiana*). Thomas Jefferson noted that it typically took one hundred of these pixies to make just a half pint. In colonial times, the soil had not yet been depleted by farming. Wild strawberries grew so thickly that hillsides took on a reddish tinge and it was impossible to take a step without crushing them. When a fire outside of Philadelphia cleared away a forest, the strawberries came on so abundantly that families traveled twenty miles to harvest them. Jefferson tried cultivating the wild berries, and swapped growing tips and plants with James Madison and the governor of Ohio.

Back in Europe, a fortuitous cross was made between the Virginia strawberry and a variety that grew along the Pacific Coast of the Americas, the Chilean strawberry (*F. chiloensis*), resulting in a dramatically larger berry. This was the pineapple strawberry (*F. ananassa*), so called because some noses detected a pineapple-like scent. It became the standard strawberry the world over, bumping the little Virginia berry from gardens and markets.

The Europeans have continued to grow their native berries, however: the intensely flavored alpine or wood strawberry (*F. vesca*) and the musk strawberry (*F. moschata*), with its musky or grapelike flavor.

Wild strawberries are not only scrumptious but also strike people as saintly in their sweetness and form. According to a quaint bit of folklore reported by Louise Beebe Wilder in *The Fragrant Path* (1932), the leaves make a tea so subtle that only those of aristocratic blood were able to enjoy them. This reverence for the strawberry goes back to the Middle Ages at least, when it was a symbol for incorruptibility and modesty. In a French miniature painting from around 1400, Joseph is seen holding out a strawberry to coax the toddler Jesus to take his first steps.

Strawberries aren't merely the pleasant smiley faces of the berry world. They have nutritional value, too, providing a good source of folic acid, potassium, and fiber. Eat just five medium-sized cultivated strawberries and you'll meet your daily requirement for vitamin C. Strawberries have been blamed for flare-ups of rheumatism and hives. If you've had problems, try switching to the alpine variety and see if the symptoms abate. A mask of crushed strawberries has been credited with fading freckles—not that anyone after Scarlett O'Hara's time should want to bother with that.

STRAWBERRIES IN THE YARD

Strawberries require some fussing. They send runners slithering out of bounds; they become fatigued and must be replanted; and they are easily overshadowed by weeds. Early frosts can nip the year's harvest in the bud. Bugs and diseases seem to find strawberries as tempting as we do. And yet the returns are more than reasonable for the investment in time.

The great number of varieties available to you can be classified by when and how they bear fruit. Traditional berries are called *June-bearing* for their habit of producing one principal crop in late spring or early summer. That means a lot of strawberries at one time, which may suit you when putting up jams or freezing large batches. You can stagger the harvest by planting early-, mid-, and late-season varieties. *Everbearing* varieties have been bred to deliver a second crop in late summer; a double harvest may be a plus, but the berries tend to be smaller and the overall yield is about the same as for June bearers. More recently, plant breeders have come up with *day-neutral* strawberries that fruit throughout the growing season. Pick them faithfully, trim back the runners, and they will perform well.

Or you might prefer to try your hand with plants from the wild, transplanting them when the plants are dormant. The Virginia strawberry puts out runners, extending from the so-called mother to its daughters. Snip the runner to a daughter and dig up this little plant with a bit of the surrounding soil, then set it out at the same depth in humus-rich garden soil. You can purchase either the red-berried plants or an unusual white-berried variation from Edible Landscaping; see "Sources," page 160.

The best place for your planting may be a patch of the vegetable garden. The area will be relatively free of weeds and chances are the soil will be reasonably fertile and loamy. There are two basic ways to grow strawberries, each involving a distinct strategy.

With the *row method,* strawberries are grown as biennials. In the first year their runners spread to cover the bare ground and the blossoms are picked to conserve the energy

CULTIVATED STRAWBERRY

that would go into fruiting. In the second year, the plants produce the crop and then are plowed under for a new, vigorous planting the following spring. So, in order to have annual crops, you'll need to maintain two plantings.

The *hill system* allows harvests over six or more years, although the plants tend to lose steam before reaching that age. They typically are set somewhat close together along raised beds. Cut back the runners as they appear, and discourage weeds and grasses from taking over the space between plants by putting down straw or sheets of black plastic. Fruits are apt to grow especially large with this method because the plants don't invest energy in producing runners. Again, blossoms are plucked the first year to get plants off to a good start, but the berries are harvested annually after that. Day-neutral and ever-bearing berries typically are grown in this fashion.

Strawberries also can be grown in pots, for the good cheer of their appearance as well as a modest yield. Gardeners of a century ago had ready access to wooden barrels, and drilled holes in them in which to tuck the plants. You can buy terra-cotta pots made for this purpose, or use rectangular planters with which to line a patio or deck. Day-neutral varieties are particularly suited to container growing. So are the European alpine and musk strawberries, both stocked by Raintree Nursery; see "Sources," page 160.

STRAWBERRIES IN THE KITCHEN

Don't pick strawberries a day or two early with the idea that they'll ripen on a sunny windowsill. They won't. Wait until their taste and fragrance are at their peak, and try to gather them early in the morning. To help prevent spoilage, put off plucking the end caps and rinsing the berries until just before using them. Keep them in the coldest part of the refrigerator for up to three or four days. Throw out any berries that show signs of mold.

Keep in mind that a given number of large strawberries will make up dramatically different volumes, depending on if and how they are processed. As a rough guideline, a 12-ounce basket of berries will come out to be 3¼ cups; slice those berries, and you're down to 2¼ cups; puree them, and you're looking at just 1½ cups.

The Virginia and alpine species tend to perform better than cultivated berries in baked goods because you take a bite and find entire berries rather than formless pink chunks.

PAN-FRIED STRAWBERRY POLENTA

In a 1642 account, colonial leader Roger Williams observed that the Native Americans of New England bruised strawberries and mixed them with cornmeal to make strawberry bread. Our Pennsylvania farm is on land where the Lenni Lenape people once lived, leaving a couple of grinding implements behind, and I wondered if they had done the same with the wild strawberries that are plentiful here in June. I tried making my own version—a polenta, using the tail ends of three bags of frozen store-bought berries. Not only was I tampering with the recipe for a traditional food, but I noticed the berries were from three continents, having been grown in the United States, Poland, and China.

Wherever they're from, strawberries bring out the sweetness in the corn. And when treated with balsamic vinegar and high heat, the berries take on interesting overtones as well.

Serves six.

3 cups water
1 cup yellow cornmeal
1 teaspoon salt
1 large red bell pepper, diced
olive oil for frying
1½ cups strawberries, diced
zest of 1 lemon and 2 tablespoons juice
2 tablespoons balsamic vinegar

Place the water in a saucepan, bring it to a boil, and stir in the cornmeal and the salt. Stirring frequently over medium heat, simmer for 15 minutes or until a sample of the cornmeal tastes done. In a skillet over medium-high heat, sauté the red pepper in olive oil. Add the strawberries and continue sautéing, allowing the berries to become somewhat caramelized by the heat. Add the lemon zest, lemon juice, and vinegar and stir well to deglaze the skillet.

Stir the pepper and strawberries into the cooked cornmeal. Spoon this mixture into a rectangular baking dish (about 9 by 13 inches) to make an even layer. Allow the polenta to sit out for 30 minutes or until it has cooled and set. Slice the polenta into six rectangles. In a lightly oiled skillet, fry the polenta until golden brown on both sides. Serve warm with a salad.

SAUTÉED STRAWBERRY SALSA

Recipes for salsa often include exotic ingredients that may be hard to find at the market and can't be grown easily in a temperate backyard. This salsa is based on strawberries, a fruit that grows north to Alaska and yet has tropical lusciousness about it. Serve the salsa with chips or try it on grilled fish and chicken. I also use it as a sandwich spread, as a topping for the morning's bowl of cheesy grits, and as an omelet filling.

Makes 1 pint.
2 tablespoons olive oil
½ red onion, chopped
1 Anaheim pepper or another relatively mild variety
2 cups strawberries, quartered
¼ teaspoon ground black pepper
¼ teaspoon salt
juice of 1 lime
1 tablespoon balsamic vinegar
2 tablespoons cilantro, chopped

In a skillet, sauté the onion and pepper in the olive oil. Add the strawberries, black pepper, and salt. Continue sautéing until the berries become caramelized but take care that they don't burn.

Ladle the mixture into a mixing bowl. While the skillet is still hot, deglaze it by adding the lime juice and vinegar and going over the bottom with a plastic spatula. Add this liquid and the chopped cilantro to the mixture in the bowl. Stir well, allow to cool, and refrigerate for use within a week.

STRAWBERRY TOMATO BARBECUE SAUCE

This summertime sauce couples subtle fruitiness with subtle heat. If you double or triple the quantities, you can freeze some for the cold-weather months. Even if you don't barbecue year-round, the sauce can be used on burgers out of a skillet or served with pasta instead of conventional tomato sauce.

Makes about 1 quart.

2 tablespoons olive oil
1 small onion, minced
2 cloves garlic, minced
2 cups fresh tomatoes, diced
2 cups strawberries, sliced
2 tablespoons balsamic vinegar
2 tablespoons maple syrup or molasses or honey
2 tablespoons bourbon
2 teaspoons hot sauce
1 teaspoon salt
1 teaspoon finely minced ginger
½ teaspoon prepared mustard
½ teaspoon Angostura bitters
zest of 1 lemon and 2 tablespoons juice
zest of 1 orange

In a saucepan, sauté the onions in the olive oil until clear. Add the garlic and sauté another minute. Add the other ingredients, simmering and stirring occasionally for 15 minutes, uncovered. Allow the sauce to cool, then pulse for 10 seconds in a food processor. Refrigerate for up to 2 weeks, or freeze to store for longer periods.

STRAWBERRY TAHINI DRESSING

At our house, the standard dressing for mesclun mixes and steamed Asian greens is made from tahini, tamari, balsamic vinegar, and olive oil, with a couple cloves of minced garlic tossed in. An aromatic variation, based on strawberries, works particularly well with a full-flavored, fifty-fifty mix of fresh spinach and arugula. This recipe is adapted from one by Drusilla Banks, nutrition and wellness educator at the University of Illinois Extension (www.urbanext.uiuc.edu).

Makes dressing for six to eight servings.
¾ cup fresh strawberries
2 tablespoons balsamic vinegar
2 teaspoons tahini
1 teaspoon tamari or ¼ teaspoon salt

Briefly pulse all the ingredients in a food processor and allow this mixture to sit for at least a half hour before tossing with the spinach-and-arugula blend.

STRAWBERRY SHORTCAKE

Recipes change from one region to another and from one dietary fashion to another. But the instructions for making an old-fashioned strawberry shortcake have remained remarkably constant. This recipe calls for some durham flour, with its coarser grind, to add a little textural interest, but feel free to use straight unbleached white flour.

Makes four or five shortcakes.
For the shortcakes:
1½ cups unbleached white flour
½ cup durham flour
½ teaspoon salt
¼ cup sugar, plus 2 tablespoons to sprinkle over shortcakes
1 teaspoon lemon zest
1¼ teaspoons baking powder
5 tablespoons butter, chilled
⅔ cup buttermilk, plus 1 tablespoon to brush on shortcakes

For the filling:

2 pints strawberries
½ cup sugar
½ teaspoon vanilla extract
1 cup whipping cream

Preheat the oven to 400 degrees F. Slice the berries thinly and place them in a bowl with ¼ cup sugar. Give the berries a stir from time to time as you make the dough. Cut the butter into dice-sized pieces and put them in the freezer 10 minutes before adding to the recipe. In a mixing bowl, use a whisk to combine the flour, salt, sugar, lemon zest, and baking powder, taking care to remove any lumps. Cut the chilled butter into this mixture using a fork, working until the texture is coarsely crumbly. Stir in the buttermilk just enough to make a workable dough. Knead the dough for about a half minute on a lightly floured surface and then form it into two same-sized balls.

Allow the dough to rest for a couple of minutes before rolling it to form two discs about ½ inch thick. To make the rounds for the shortcakes, use a 3-inch-diameter cookie cutter, or run a knife around the bottom of a can of roughly that size. Cut four or five rounds from each disc of dough, gathering the scraps and rolling them again if necessary. Brush the top surface of the shortcakes with the 1 tablespoon of buttermilk and sprinkle with the 2 tablespoons of sugar. Place on a baking sheet and bake for 15 to 20 minutes, or until slightly golden.

Soften the berries with a potato masher or the back of a fork. Combine the cream, sugar, and vanilla extract in a mixing bowl and whip this mixture to form peaks. Take the rounds from the oven and assemble the shortcakes with their brushed tops facing up. Spoon the berry mixture onto the bottom rounds, put the top rounds in place, spoon on more berries, and crown with the whipped cream.

STRAWBERRY SCONES

As with bagels and focaccia, store-bought scones can look appealing but offer little in the way of taste and texture. This recipe uses both baking powder and chilled butter for a light, flaky, biscuitlike texture. The scones are assembled like a sandwich, containing a layer of strawberry jam. The basic procedure isn't much different from that in the strawberry shortcake recipe, above.

Makes twelve scones.

2 tablespoons butter (for top of scones)
7 tablespoons butter (for the dough)
3 cups unbleached white flour
⅓ cup sugar
2 teaspoons baking powder
½ teaspoon baking soda
1 teaspoon salt
1 cup buttermilk
¼ cup strawberry jam

Preheat the oven to 425 degrees F. Cut the 7 tablespoons of butter into dice-sized pieces and put them in the freezer 10 minutes before adding to the recipe. Use a whisk to combine the flour, sugar, baking powder and soda, and salt, taking care to remove any lumps. Cut the chilled butter into this mixture using a fork, working until the texture is coarsely crumbly. Pour in the buttermilk. Again using the fork, mix the dough just enough to combine the ingredients well.

Add a little buttermilk or water if necessary to form the dough into two same-sized balls. Place each ball in turn on a floured work surface and press with the heels of your hands to flatten it somewhat. Continue pressing or use a roller to make a 7-inch circle. Spread the jam over one of the circles, going not quite to the edge. Put the other circle on top, and pinch lightly around the edge to contain the jam. Roll this circle until it is about 12 inches across.

Melt the 2 tablespoons of butter. With a knife, make marks around the perimeter of the circle to guide you in cutting 12 equal-sized wedges. (Think of where the hours are on a clock face). Use the knife to cut the wedges. Place them on an ungreased cookie sheet and brush the tops with the melted butter. Bake for 12 to 20 minutes or until the tops just turn golden brown.

STRAWBERRY RIESLING DESSERT SOUP

This soup comes together without much fuss. Frozen strawberries will work, but use fresh ones if possible.

Serves four.

1 pint fresh or frozen strawberries, halved, plus several sliced as garnish
½ cup Riesling wine
1 tablespoon Grand Marnier or Cointreau
zest of 1 lemon
⅓ cup sugar

Put the halved strawberries in a mixing bowl and add the other ingredients, stirring to mix well. Allow this mixture to sit for 30 minutes or so, then puree it in a food processor. Refrigerate the soup until you are ready to serve it, then garnish each portion with sliced berries.

STRAWBERRY RHUBARB TAPIOCA

Strawberries and rhubarb appear in the garden (and in supermarkets) about the same time each year, and they make a good sweet-tart pair in all sorts of recipes. Lyla Derr grows rhubarb at her Pennsylvania home and uses this pudding recipe from her daughter-in-law in nearby Pennsylvania Dutch country, where sweet pies and puddings are legend.

Serves six.

1 quart chopped rhubarb

1 quart fresh strawberries, halved,
plus 6 whole berries as garnish

¾ cup sugar

¼ cup minute tapioca

¾ cup water

1 cup whipping cream (optional)

Soak the tapioca in cool water for 30 minutes. In a saucepan, cook the soaked tapioca, rhubarb, and sugar over medium-high heat, stirring until the mixture reaches a boil. Turn the heat to low and continue cooking until the mixture is transparent. Gently stir in the strawberries and allow the pudding to cool. The flavor seems to improve after a day in the refrigerator. Top each serving with whipping cream, if you wish, and a whole berry.

STRAWBERRIES ROMANOFF

This simple recipe with the exotic name has been knocking around North American kitchens at least since the 1930s. Serve it in wine or cocktail glasses and it will look like more than the sum of its few ingredients.

Serves six.

2 cups fresh strawberries, halved

⅓ cup Grand Marnier or Cointreau
(or triple sec plus zest of 1 orange)

½ cup whipping cream

½ teaspoon vanilla extract

2 tablespoons confectioners' sugar

If you're using mammoth strawberries, cut them down to bite-sized pieces; otherwise, cutting them in half will do. Marinate the berries for a half hour in the liqueur. (To save money and halve the alcoholic content, substitute triple sec and orange zest for a roughly equivalent taste.) Puree one-third of the marinated berries. Whip the cream with the vanilla extract and sugar. Fold the puree into the whipped cream. Divide the remaining berries between four glasses, and top with the mixture of whipped cream and puree.

STRAWBERRY FUNNEL CAKE

This recipe might serve as the cornerstone of a healthful breakfast if it weren't for one ingredient—a shallow sea of vegetable oil. Funnel cake, in case you haven't made its acquaintance, is a confection of fried dough most often savored at carnivals and country fairs. The batter is poured through a funnel in order to draw lacy patterns upon a skillet of hot oil. As greasily delectable as the commercial variety may be, this one is better. The strawberries and real vanilla extract put off an irresistible scent when the batter hits the oil. Safflower is a good choice for the oil—better for your health than certain other oils, and to my nose it has a fresher scent when heated than canola.

A word of caution: although funnel cakes are a great treat for kids, younger children should not help make them because of the danger when cooking with hot oil.

Yield varies with skillet size and design of cakes.

1½ cups fresh or frozen strawberries
¼ cup sugar
½ cup water
¾ cup low-fat milk
½ teaspoon vanilla extract
whites of 2 eggs
2 cups unbleached white flour
½ teaspoon salt
1¼ teaspoon baking powder
vegetable oil to fry
confectioners' sugar

Place the strawberries, sugar, and water in a saucepan and simmer for 10 minutes, stirring occasionally. Puree this mixture in a food processor until smooth; clumps of berry are apt to stop up the funnel. Add the milk and vanilla extract. Whisk the egg whites and add them. In a mixing bowl, thoroughly combine the flour, salt, and baking powder. Add roughly *two-thirds* of these dry ingredients to the berry mixture, and beat to make a smooth batter. Use the reserved third to thicken the batter if necessary, or add a tablespoon of milk at a time if the batter is too thick. The consistency should be that of pancake batter. Do a trial run by pouring ¼ cup of batter through the funnel and back into the bowl. If the batter is too thick (like frosting), it won't flow properly through the funnel; if too runny, it will spread uncontrollably over the oil.

Pour the vegetable oil in a skillet 6 to 9 inches in diameter, to a depth of just over ¼ inch. (The larger the skillet, the larger and more elaborate you can make the funnel cakes, but you'll need proportionately more oil.) Heat the oil to 375 degrees F, as measured by a cooking thermometer. If you don't have a thermometer, try dropping a half teaspoon of batter into the oil. It is hot enough if the batter browns nicely in 5 seconds or so. If the oil begins to smoke, turn down the heat.

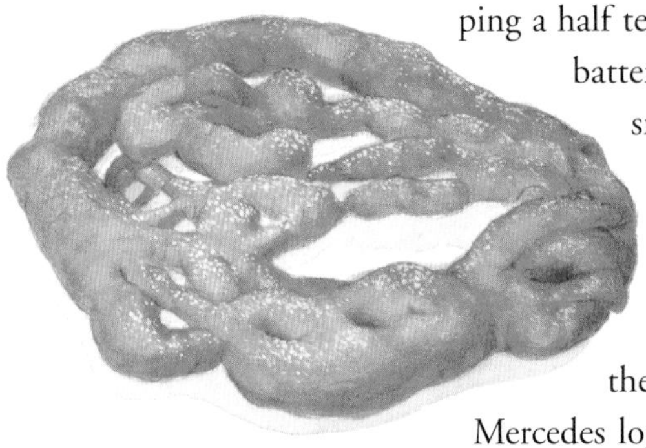

Pour ¼ cup of batter into a funnel held in one hand, keeping a finger over the end as a valve. Move the funnel slowly over the skillet to make a pattern in the oil. Begin with a circle around the perimeter, then make cross strands similar to a peace sign or Mercedes logo. Fry the cake for about 20 seconds on each side, turning it with tongs and then removing it to a couple thicknesses of paper towel placed on a cake rack. Sprinkle with confectioners' sugar if you wish. Once they've drained a minute, the cakes can be kept warm in a toaster oven. But eat them soon—they aren't any better when cold than hour-old french fries. Pour the leftover oil through a strainer for reuse.

CRÈME DE FRAISE

Crème de fraise is a highly aromatic strawberry liqueur with a fresh-from-the-garden taste. To make your own, half fill a glass jar with sliced berries, then pour in vodka nearly to the top. Put on the lid and place the jar in a cool, dark place for two or three weeks, giving an occasional shake. Vivi Labo, who makes various liqueurs and flavored vodkas in her Danish home, is careful to use a high-proof vodka when macerating large cultivated strawberries. That's because their high water content interferes with the alcohol's ability to absorb the berries' essence. Strain through cheesecloth placed in a colander, and again through a coffee filter for greater clarity. To turn this flavored vodka into a liqueur, sweeten to taste with simple syrup (see the recipe on page 21). Pour the liqueur into bottles. You can begin enjoying it right away or age it for later use.

MULBERRY

CHAPTER 13

Mulberries

It is difficult to avoid noticing the enormous dark berry stains left under mulberry trees, as well as on drying laundry and automobiles, thanks to berry-eating birds. And yet the generous harvest is produced without much fanfare, save for a dedicated few people to whom baking mulberry pies is a summer tradition. Mulberries usually go unappreciated. They're the fruit that looks something like a blackberry but lacks its personality.

Of the three main varieties of mulberry growing in the United States, the red (*Morus rubra*) is native, its territory stretching from New England south to Florida and west to Oklahoma and Texas. The flavor of the fruit (they aren't really berries) is rated second best to the imported black mulberry (*M. nigra*), varieties of which can be purchased from nurseries specializing in fruiting plants. Then there is the white mulberry (*M. alba*), with its pale, insipid-tasting fruit. In a visionary scheme, white mulberry trees were brought here by the thousands as a food source for silk-producing worms. The silkworm boom went bust in 1839, but the white mulberry itself became established in much of the country. The trees do too well, in fact, and now rank as the second most common weed tree in New York City.

MULBERRIES IN THE YARD

Given the right conditions (and they aren't a demanding tree), mulberries sprout up everywhere. Just don't place them where the falling fruit will be a problem—over patios and driveways or near the clothesline. Fruit nurseries offer several varieties with especially good flavor, including Illinois Everbearing and Oscar. The unusual weeping mulberry forms a small outdoor room that can accommodate a few adults at teatime or serve as hideout for the kids. Raintree Nursery sells this and other mulberries by mail order; see "Sources," page 160.

MULBERRIES IN THE KITCHEN

To harvest the fruits, you don't need a stepladder. Just place a clean sheet beneath a branch and give it a good shake. Easier still, allow the berries to plop down on their own, but you'll have to carefully pick out sticks, leaves, and other debris. Because they are so fragile, mulberries should be either used as soon as possible or frozen.

Mulberries can serve in recipes calling for blackberries or raspberries, with one caveat—they are best combined with other berries or apples or pears. A hundred years ago, gardening magazines suggested readers grow red currants to add to mulberry pies, and that's still good advice. Not that everyone thinks mulberries need correction. A huge mulberry tree grows by the tennis courts I used to frequent, and I could count on seeing families wade through the weeds to reach it each summer when the fruit began to drop.

"They're kind of bland, aren't they?" I asked an old man, hoping he'd share a recipe.

"No they're not," he said, sounding offended, and that was the end of our exchange.

MULBERRY RHUBARB PIE

That uniquely tart vegetable, rhubarb, wakes up the mulberries in this simple pie. Either buy a ready-made crust or make the one for Wild Blueberry Pie on page 68.

Serves six.

2½ cups mulberries
1½ cups rhubarb, chopped finely
1¼ cups white sugar
¼ cup flour
2 teaspoons tapioca
2 tablespoons butter

Preheat the oven to 425 degrees F. In a mixing bowl, stir together the mulberries, rhubarb, sugar, flour, and tapioca. Allow this mixture to stand for 10 minutes. Pour them into the pie crust, and place pats of butter over the surface. Put the top crust in place. Bake initially at 425 degrees for 10 minutes, then turn the oven down to 375 degrees and cook for another 35 minutes or until the crust shows good color.

CHAPTER 14

Juniper Berries

The scent of juniper is haunting and difficult to describe. When simmered in a sauce or milled finely in a coffee grinder, the berries put off the clean scent of a workshop where cedar boards are being milled. Less obvious is a hint of citrus, perhaps the peel of an orange.

As with a number of nose-pinching herbs and spices, juniper berries long have been considered to possess healthful properties. In his book *Chymistry Applied to Agriculture,* published in 1836, the French author John Antony Chaptal declared that a wine of juniper berries was "one of the most wholesome" drinks for farm laborers, even though "it requires a little use to reconcile one to the odor and flavor of it." Many generations

JUNIPER BERRY

later, the gin martini has a reputation as a bracing (if not overtly healthful) drink that also takes getting used to.

In cooking, the berries are used as a spice, particularly to make game and wild meats a little less . . . well, gamey. Quail, duck, venison, and boar are complemented by a sauce or marinade prepared with juniper berries. Try them in full-bodied vegetable dishes, too, such as baked beans, hearty soups, spiced red cabbage, sauerkraut, and a mash of potatoes with turnips or rutabagas.

This particular juniper is an evergreen shrub, *Juniperus communalis,* that grows in the wild in the Northern Hemisphere. You sometimes can find the dried berries packaged in a supermarket or specialty store. They are leathery and look a bit like hardened black currants or blueberries. In fact, they aren't berries at all but can be described more accurately as seed cones. You can grow your own; search local nurseries or mail-order sources specifically for *Juniperus communalis* and not just any landscaping plant identified as a juniper.

JUNIPER BERRIES IN THE KITCHEN

Once picked or purchased, the berries are best used within a year or two because their essence is gradually lost.

Whole juniper berries are no more a treat to bite into than peppercorns. You can remove them from dishes before serving. Or simmer the berries in a bit of water or wine, strain, and add the liquid to the food. Another option is to pulverize the berries. An electric coffee grinder works well. You also can crush the berries with a mortar and pestle, or mince them finely with a knife, but the resulting particles will be more obvious in the dish. To get the flavor of juniper without the hassle, use a couple of tablespoons of gin instead of the berries.

BALSAMIC JUNIPER GLAZE

A friend, Gail Stern, likes to stir up a juniper berry marinade when broiling salmon. The molasses thickens the marinade so that it doesn't immediately roll off the fish. Then, under the broiler coil of an oven, the heat caramelizes the marinade into an attractive glaze with the fresh scent of the berries.

Yields about ½ cup.

1 rounded teaspoon juniper berries
6 tablespoons balsamic vinegar
4 tablespoons molasses
½ teaspoon salt

Grind the berries as described earlier in the section "Juniper Berries in the Kitchen." In a small saucepan, heat the vinegar to a simmering boil, stir in the molasses, then add the ground berries. Simmer for 5 minutes. Pour the mixture into a shallow dish or mixing bowl and rest the fish in it, turning to treat both sides.

GIN AND JUNIPER CHICKEN

Roasted and grilled birds seem to go well with the piney tang of juniper. This recipe, based on one from the Gin and Vodka Association in the United Kingdom, uses both gin and juniper berries, along with a nudge of red wine.

Serves four.

4 tablespoons butter
4 medium-sized chicken breasts
1 medium onion, diced
1 clove garlic, crushed
1 rounded tablespoon chopped finely fresh parsley
1 tablespoon prepared mustard
1 cup chicken or vegetable stock
½ cup dry red wine
4 tablespoons gin
1 rounded tablespoon juniper berries
1 cup basmati rice, with water to prepare

Preheat the oven to 350 degrees F. In a saucepan, sauté the chicken in 2 tablespoons of butter until browned. Transfer the chicken to an ovenproof dish with a cover. Again in the saucepan, sauté the onions in the remaining tablespoons of butter until soft. Puree the garlic, parsley, mustard, stock, wine, gin, and juniper berries in a food processor, making sure the berries are reduced to small fragments. Add this mixture to the onions and combine well, then pour the sauce over the chicken and cover. Bake for 30 to 40 minutes. Serve with the rice.

TEPARY BEANS WITH JUNIPER AND ROSEMARY

Beans are a foursquare food that can be enlivened with an aromatic hit of juniper. Any sort of bean will do for this recipe, but teparies have a nutty and subtly sweet flavor. Teparies are a Southwestern bean, suited to the dry climate and grown for centuries by the Native Americans of the region. You can find different varieties of tepary beans through mail order; one source is Native Seeds/Search, 526 North Fourth Avenue, Tucson, AZ 85705-8450.

Serves four as a side dish.

1 cup dry tepary beans (or other beans)
water to cook beans
8 juniper berries
2 teaspoons fresh rosemary, minced
½ small onion, minced
½ teaspoon hot sauce
½ teaspoon salt
4 tablespoons olive oil
juice of 1 lemon
½ bunch fresh parsley, chopped

Rinse the beans and examine for stones. Place the beans in a saucepan, cover with water, and simmer over low heat. To reduce the effect of beans on the digestive system, discard the water after 15 minutes or so of cooking, cover with fresh water, and continue to simmer. Give an occasional stir, and add water as necessary.

In the meantime, grind the berries as described earlier in the section "Juniper Berries in the Kitchen." Place the olive oil in a skillet. Sauté the berries, rosemary, onion, salt, and pepper over medium-low heat for 10 minutes.

Continue cooking the beans until they are tender but not yet mushy in texture. Drain the beans and place them in a mixing bowl. Add the sautéed mixture, the lemon juice, and the parsley and stir well. The beans will be at their most fragrant if served while still warm.

JUNIPER ORANGE PANCAKES

The juniper and orange give these pancakes a Christmas-like flavor, and the heartiness of the cakes is suited to winter breakfasts. Serve with the accompaniments of your choice: yogurt or sour cream; maple or berry syrup; or applesauce.

Serves four.
½ cup milk
12 juniper berries
¼ teaspoon ground black pepper
zest of 1 orange
½ teaspoon salt
1 egg (or whites of 2 eggs)
¼ cup unbleached white flour
2 large potatoes (about 1 pound), grated
olive oil for cooking

Pour the milk into a mixing bowl. Grind the berries as described earlier in the section "Juniper Berries in the Kitchen." Add the ground berries, pepper, orange zest, salt, and egg to the milk. Whisk well. Stir in the flour. Grate the potatoes and stir them into the batter. Spoon thin cakes onto a lightly oiled skillet and cook over medium heat until golden brown on both sides. Serve hot.

JUNIPER PUMPKIN PIE WITH BERRY CREAM

In pumpkin pie, spices are the thing. This recipe skips the customary cinnamon because it obscures other flavors and scents. Instead, you'll use juniper berries buffered with a bit of gin. The result is subtle, with the juniper and gin remaining very much in the background. Add flavor and color to the whipped cream with a bit of any berry syrup or liqueur you have on hand.

Serves six.

For the pie:

1 two-cup graham-cracker-crust pie shell
1¾ cups butternut squash, without the rind
1 level tablespoon juniper berries
¼ cup water
1 whole egg and white of 1 egg
1¼ cups condensed milk
¼ cup dark brown sugar
2 tablespoons honey
1 teaspoon grated fresh ginger
½ teaspoon grated mace
½ teaspoon grated nutmeg
½ teaspoon ground black pepper
½ teaspooon salt
3 tablespoons gin

For the cream:

½ pint heavy cream
¼ cup berry syrup or liqueur

Preheat the oven to 425 degrees F. With a heavy knife or cleaver, chop the squash into several pieces and place them in a covered skillet with 1 inch of water. Cook over medium heat for about 10 minutes or until the squash is softened. Grind the berries as described earlier in the section "Juniper Berries in the Kitchen." Simmer them in ¼ cup water for 5 minutes, then strain the aromatic liquid through a coffee filter. Allow the squash to cool so that you can handle it, then cut the flesh away from the rind and puree it in a food processor. Whisk the eggs in another bowl, and add them to the squash along with the juniper liquid, milk, sugar, honey, ginger, mace, nutmeg, salt, pepper, and gin. Pour this mixture into the crust.

Bake for 15 to 20 minutes, then turn down the heat to 350 and bake for an additional 25 to 35 minutes, or until the filling is firm. Whip the cream until stiff, then stir in the berry syrup or liqueur. Allow the pie to cool somewhat before serving slices with the whipped cream.

BATHTUB GIN

Gin is a complex concoction of fresh and dried botanicals in which juniper berries play a key role. Other commonly used flavorings include lemon and orange zest, cardamom, caraway, coriander, fennel, and licorice root.

For a simple homemade gin, crush a rounded tablespoon of juniper berries and put them in a jar with 20 ounces of good vodka. Add small amounts of other traditional gin ingredients if you wish to experiment. Set the jar aside in a cool, dark place for from four days to two weeks, and give it an occasional shake. Strain into a bottle, seal, and age for at least a month. You may find that oil globules form on the surface; simply shake the gin before serving and they will be inconspicuous. If the taste is too strong and spicy, dilute the gin with vodka.

ACKNOWLEDGMENTS

Thanks to all of the berry lovers who shared recipes and anecdotes with me. I am particularly indebted to my wife Ali for harvesting and freezing berries, testing recipes, and finding the candy thermometer in the pile of old place mats. In Ed Mashburn I came to know the Johnny Appleseed of unsung berries, and I was inspired by his hard work out in the sultry, buggy rows of trial currants and gooseberries. Finally, this book wouldn't have seen print if it weren't for Jim Mairs, my editor at W. W. Norton and a champion of *all* things unsung.

SOURCES

Edible Landscaping, 361 Spirit Ridge Lane, Afton, VA 22920
Forest Farm, 990 Tetherow Road, Williams, OR 97544
Hartmann's Plant Company, P.O. Box 100, Lacota, MI, 49063-0100
Nourse Farms, Inc., 41 River Road, South Deerfield, MA 01373
Raintree Nursery, 391 Butts Road, Morton, WA 98356

ALSO BY THE AUTHOR

Apples
A Celebration of Heirloom Vegetables

FOR YOUNG READERS:

City Trains
Humanpower
Smarten Up!
Train Talk

This book is composed in Adobe Garamond
Book design and composition by Katy Homans
Manufacturing by Mondadori Printing, Verona, Italy